STUDIES IN GLOBAL GEOMETRY AND ANALYSIS

Studies in Mathematics

The Mathematical Association of America

Charles W. Curtis, editor

Marston Morse
Institute for Advanced Study, Princeton

S. S. Chern
University of California, Berkeley

Harley Flanders
Purdue University

Shoshichi Kobayashi
University of California, Berkeley

Lamberto Cesari
University of Michigan

L. A. Santalo
University of Buenos Aires

Studies in Mathematics

Volume 4

STUDIES IN GLOBAL GEOMETRY AND ANALYSIS

S. S. Chern, editor
University of California, Berkeley

Published and Distributed by
The Mathematical Association of America

© 1967 by
The Mathematical Association of America (Incorporated)
Library of Congress Catalog Card Number 67-16033

Complete Set   ISBN 0-88385-100-8
Vol. 4   ISBN 0-88385-104-0

Printed in the United States of America

Current Printing (last digit):
10   9   8   7   6   5   4   3

CONTENTS

INTRODUCTION

S. S. Chern

Global geometry or analysis is a natural outgrowth of the infinitesimal calculus. While the calculus deals with the theory of differentiable functions in one or more real variables, modern differential geometry or analysis centers its attention on manifolds. Manifolds are spaces which behave locally like euclidean spaces and to which the notions of differential and integral calculus can be extended.

However, such an extension is elaborate, and its final form has perhaps not been achieved. The differential aspect is essentially the tensor calculus, with its modern ramifications. The generalization of integration proceeds in two different directions: measure theory and an algebraic theory which leads to exterior differential forms and cohomology theory in algebraic topology. The importance of manifolds in modern mathematics cannot be overemphasized.

From the theory of manifolds arises the notion of a fiber space. A fiber space behaves locally like a Cartesian product and usually

1

carries further algebraic structures, such as a vector space structure, in the fiber. It is perhaps fair to say that the development of the global theory of fiber spaces, in its various ramifications, includes a substantial amount of the mathematical work done in the last two decades.

This volume, as others in this series, contains articles on the upper undergraduate and graduate levels. They are supplementary readings and thus do not contain material usually covered in courses or standard textbooks. Nevertheless, the problems treated are lively and indicate the areas of active, recent work.

Professor Morse's paper is a reprint of his expository article in the *American Mathematical Monthly* in 1942, with a new section added on the latest literature. After nearly 25 years, the paper is as lively as ever. Very few ideas in mathematics are as uncomplex and as powerful as the Morse critical point theory. In its simplest form, this theory says that the behavior of a smooth function on a compact manifold at its critical points is strongly restricted by the topological properties of the manifold, or (which is the same) that much information on the topology of a manifold can be obtained from its "nice" smooth functions. The critical point theory extends to infinite-dimensional manifolds, and, among other consequences, leads to conclusions on extremals or closed extremals of variational problems. Applications of Morse's theory are unlimited.

The second paper, by this writer, is concerned with some examples of global differential geometry in euclidean space. Although we all favor equal rights for manifolds, the euclidean space, with its intuitive background, occupies a unique position. Theorems on the geometry in euclidean space are attractive because they are easily understood. The elementary nature of euclidean space is, however, misleading. We can ask any number of questions on global surface theory in euclidean space, the answers to which are unknown. Hopefully, some theorems in this paper will be the forerunners of more general results on the isometric mappings of Riemannian manifolds.

Professor Flanders' paper is an elementary account on exterior differential forms which could be considered an introduction to his beautiful book (H. Flanders, *Differential Forms With Applications to the Physical Sciences*. New York: Academic Press Inc., 1963). The exterior differential calculus was formalized by the great geometer Élie Cartan, who exploited it with remarkable success. Of course, special cases were known long before: The differential forms are but the integrands of multiple integrals, while exterior differentiation generalizes the notions of the gradient and the curl in vector analysis and gives the algebraic content of Stokes's theorem. The importance of the exterior differential calculus lies in its simplicity. It is one of the few operations which can be defined on any differentiable manifold without the use of an additional structure (such as a Riemannian metric or an affine connection), and it is a natural operation, in the sense that it commutes with differentiable mappings. The topological interpretation of the exterior differentiation as a coboundary operator and the exterior multiplication as the cup product add luster to an already brilliant star.

Professor Kobayashi's paper is probably more advanced than the three just discussed. It is, however, intended as an introduction to an important topic in Riemannian geometry and the treatment is elementary. The conjugate and cut loci of a Riemannian manifold are geometrically very suggestive, and their properties should reflect much of the geometry on the manifold itself. They already played an essential role in the so-called pinching theorems on Riemannian manifolds of positive curvature (H. Rauch, W. Klingenberg, M. Berger, D. Gromoll, E. Calabi, etc.). These pinching theorems are among the major achievements in Riemannian geometry in the last decade. They pertain to the general question on the study of Riemannian manifolds of positive sectional curvature. A classical result says that a compact orientable two-dimensional manifold of positive curvature can be isometrically realized as a convex surface in euclidean 3-space (Weyl's problem). In higher dimensions, the situation is much more involved: Probably few compact simply connected Riemannian manifolds of positive curvature exist, but our knowledge of them is extremely limited.

The papers of Professors Cesari and Santalo both concern integration as measure, but their difference in spirit offers an interesting contrast. While Professor Cesari gives a rigorous foundation of the theory of area based on his own work, Professor Santalo's paper contains a wealth of information on relations between integral invariants. Integral geometry plays a role in many geometrical problems. A recent development is the generalization of the Radon transform, which has important applications to partial differential equations and group representations (Gelfand and his school and Helgason).

Before concluding this introduction, I would like to mention briefly two of the most important recent developments. The first is the bordism theory of R. Thom. As is the case for all important mathematical concepts, Thom's basic idea is extremely simple. Two compact oriented manifolds of the same dimension are said to be bordant if their difference is the boundary of a manifold of one higher dimension. These "bordism classes" can be made to form a ring of a determined structure, which, of course, gives deep insight into the family of all compact manifolds.

The second development can be described as the search of relations between analytical and topological invariants. An example is the classical Riemann-Roch theorem on a compact Riemann surface, which expresses in terms of the genus of the surface and other invariants the dimension of the vector space of all meromorphic functions with poles and zeros that satisfy certain relations. A far-reaching generalization applies to elliptic operators on sections of vector bundles over a compact manifold and to the study of their "indices" (M. Atiyah, R. Bott, F. Hirzebruch, and I. Singer). (See F. Hirzebruch, *Topological Methods in Algebraic Geometry*, New York: Springer, 1966, and R. S. Palais, "Seminar on the Atiyah-Singer index theorem," Princeton, 1965.) This generalization requires, among other things, the consideration of singular integral operators on manifolds (A. Calderon, A. Zygmund, and R. Seeley). Geometry, topology, and analysis are now indistinguishable.

WHAT IS ANALYSIS IN THE LARGE?

Marston Morse

1. INTRODUCTION

All mathematics is more or less "in the large" or "in the small." It is highly improbable that any definition of these terms could be given that would be satisfactory to all mathematicians. Nor does it seem necessary or even desirable that hard and fast definitions be given. The German terms *im Grossen* and *im Kleinen* have been used for some time with varying meanings. It will perhaps be interesting and useful to the reader to approach the subject historically by way of examples.

No proofs are given. In attempting to give the reader a conception of analysis in the large two ways are open. The first is to attempt an elementary exposition of the fundamental techniques. Unfortunately, this method of exposition is attempted much too often. The explanations given are fragmentary and give an exaggerated notion of the importance of some special technique, and no adequate notion of the subject as a whole. In a new and comprehensive field possibly the only way to give the beginner a stimulating and adequate notion of what the subject is about is to give examples and results which are themselves relatively complete. The cooperative reader can readily imagine the variety of techniques that might be used to obtain the stated results, and may himself invent new techniques, but in the presence of significant results he is less apt to be concerned with trivialities and subjective bypaths.

2. AN EXAMPLE FROM DIFFERENTIAL GEOMETRY

Most of classical differential geometry is "in the small," that is, most theorems are proved merely in the neighborhood of a point. It is proved, for example, that in the neighborhood of a point P of a surface Σ, Σ can be referred to isothermic parameters so that neighboring P,

$$(2.1) \qquad\qquad ds^2 = \lambda(u, v)[du^2 + dv^2]$$

with $\lambda(u, v) \neq 0$.† The question in the large as to what sort of closed surfaces can be represented as a whole with parameters (u, v) and ds^2 of the form (2.1) has been asked and answered in general only in recent years. It is required that there be just one curve $u = $ const. and just one curve $v = $ const. through each point. Among two-sided or orientable surfaces which admit such parameters, those of the topological type‡ of the torus are the only possibilities.

One could continue by asking a more general question. What sort of closed surfaces S admit a representation in terms of parameters (u, v) in such a manner that there is one and only one curve $u = $ const. and one and only one curve $v = $ const. through each point? Such a representation of S would in particular imply the existence at each point P of S of a vector tangent to the curve $u = $ const. through P. There would thus exist a field of vectors,

---

† Appropriate hypotheses as to the regularity of the representation of the surface must be made.

‡ A surface is of the topological type of the torus if it is the (1–1) continuous image of the torus.

one for each point of P, tangent to S and P and varying continuously with P. For such a field to exist S must be the topological type of the torus.

Thus, in questions as to the existence of parameter nets without singularities, the controlling factors are those of topology. One can see why analysis or geometry in the large depends so heavily on topology.

3. AN EXAMPLE FROM THE THEORY OF
FUNCTIONS OF A COMPLEX VARIABLE

The theorem that a function $f(z)$ of a complex variable z which has no singularities in the extended plane other than poles is a rational function of z, is a theorem in the large the proof of which illustrates some of the salient characteristics of analysis in the large. One begins by representing $f(z)$ neighboring $z = z_0$ as the sum of the "principal part" of $f(z)$ at z_0 and a function analytic at z_0. This is the preliminary analysis in the small.

Upon subtracting the principal parts of $f(z)$ at each pole from $f(z)$ one obtains a function $\phi(z)$ bounded in absolute value and with at most removable singularities. According to Liouville, $\phi(z)$ is a constant. The theorem follows.

The analysis in the large comes in the proper definition of the extended plane and the proof of the Liouville theorem. Details will not be given but it will be of interest to state that the theorem of Liouville can be reduced to a theorem of topological character on the nature of vector fields.

4. DIFFERENTIAL EQUATIONS IN THE LARGE.
AN EXAMPLE FROM THE WORKS
OF HENRI POINCARÉ

It is no mere coincidence that Poincaré was the first to comprehend fully the possibilities of analysis in the large, and at the same time was the father of modern topology. Poincaré was not satisfied with the classical theory of differential equations. He wished to know something concerning the system of trajectories as a whole. He was greatly interested in the movements of the planets but found insufficient generality and completeness in the classical theory. His interest in celestial mechanics is in the background of all of his papers on differential equations.

Poincaré's first papers on differential equations are not pretentious in their generality, but in method they are most novel. Poincaré is concerned with an ordinary first order differential equation[†] defined at each point of a 2-sphere. In terms of any system of local coordinates (u, v) representing the neighborhood of a point (u_0, v_0) on the sphere the differential conditions have the form

$$\frac{du}{U(u, v)} = \frac{dv}{V(u, v)}.$$

The functions U and V are supposed real and analytic in (u, v) neighboring (u_0, v_0). Points (u_0, v_0) at which both U and V vanish are termed "singular points." These points are supposed finite in number on the sphere.

Poincaré makes certain assumptions concerning the singular points (u_0, v_0). To state these conditions we shall take (u_0, v_0) as the origin. Then U and V have developments of the form

$$U = au + bv + \cdots,$$
$$V = cu + dv + \cdots$$

neighboring the origin. Poincaré assumes in most of his work that the roots λ_1 and λ_2 of the equation

$$\begin{vmatrix} a - \lambda, & b \\ c & d - \lambda \end{vmatrix} = 0$$

are distinct, different from 0, never pure imaginary, and that neither $\lambda_1|\lambda_2$ nor $\lambda_2|\lambda_1$ is a positive integer. These conditions will be satisfied by most analytic examples.

Curves on the sphere which satisfy the differential equation are termed *characteristics*. In general, characteristics are without singularity except at most when they pass through a singular point

---

[†] Poincaré, "Sur les courbes définies par les équations différentielles," *Journal de Liouville*, 1881, 1882.

of the differential equation. Typical of analysis in the large, Poincaré's work permits a subdivision into three parts as follows:

(a) *a study of characteristics neighboring a singular point;*

(b) *the assignment of an index* ±1 *to each singular point and the establishment of a relation between these indices* (*this part of the analysis would now be regarded as an essay in combinatorial topology*);

(c) *a description of the characteristics in the large with particular reference to recurrence and limiting trajectories* [*results* (a) *and* (b) *are preliminary to* (c)].

Part (a). In his study of characteristics neighboring a singular point, Poincaré shows that there are three principal kinds of singular points as follows:

"*Noeud.*" Neighboring a noeud (u_0, v_0) each characteristic tends to (u_0, v_0) with a definite limiting direction. For example, the differential equation

$$\frac{du}{u} = \frac{dv}{2v}$$

has a noeud at the origin. In this example, the characteristics have the form $kv = hu^2$ where h and k are constants.

"*Foyer.*" The characteristics approach such a singular point in the form of spirals, with the arc length becoming infinite. For example, the differential equation

$$\frac{du}{u - v} = \frac{dv}{u + v}$$

has a foyer at the origin with logarithmic spirals as characteristics.

"*Col.*" There are just two characteristics which tend to a col as a limiting point. For example, the equation

$$\frac{du}{u} = \frac{dv}{-v}$$

has a col at the origin. The characteristics $uv = $ const. include the two characteristics $u = 0$ and $v = 0$ passing through the origin.

Part (b). In the development (b), Poincaré assigns an index 1 to each noeud and to each foyer, and an index -1 to each col.

Poincaré shows that *the sum of the indices of the singular points on the sphere equals* 2. Thus, there must exist at least two singular points.

A closed characteristic without a multiple point is called a *cycle*. If a characteristic tends to a noeud or a foyer as a limit point there is in general no natural way to continue the characteristic, and it is agreed that in such cases the characteristic shall end at the noeud or foyer. If a characteristic *g* tends to a col the convention is made that *g* may be continued turning either to the right or left and departing from the col on a characteristic. By virtue of this convention, the notion of a cycle is enlarged. With this understood we see that a cycle can have no singularity other than those occurring at a col.

Part (c). Poincaré ends with a relatively complete description of the characteristics. He shows that *a characteristic continued without limit in a given sense either terminates at a noeud, or is a cycle, or is asymptotic to a cycle.* A foyer is to be regarded as a degenerate cycle to which the neighboring spirals are asymptotic.

The reader is asked to observe the fundamental difference between the modes of analysis required in Parts (a), (b), and (c), and then to note how (a) and (b) are preliminary to (c) and make (c) possible. The index theorem of Poincaré has its topological generalization in the fixed point theorems of Brouwer, Alexander, Lefschetz, and H. Hopf. The analysis of characteristics in (c) is the predecessor of the modern study of recurrence and transitivity which G. D. Birkhoff has developed so fully and to which Hedlund, Morse, von Neumann, Koopman, E. Hopf and others have made significant contributions.

5. ELEMENTARY EXAMPLES IN EQUILIBRIUM THEORY IN THE LARGE

Equilibrium theory in the large makes an extensive use of topology. The principles of analysis brought out in the previous examples appear here again. Briefly summed we have seen in these examples that analysis in the large has involved (a) a preliminary analysis in the small, (b) a local determination of indices, and (c) an integration of this local analysis by various means (including topology) into the final theorems in the large. The examples which we shall now present will show how various problems which from a local point of view appear most diverse, from a topological point of view are essentially the same.

We begin with certain results concerning a function f of a point on a closed bounded n-manifold Σ lying in a euclidean space of sufficiently high dimension. We suppose throughout that Σ is locally represented in terms of n parameters (u) with convenient conditions of differentiability and regularity. In terms of the local parameters (u) f shall be a function $F(u)$ at least three times continuously differentiable. A *critical* or *equilibrium point* of f is a point at which each partial derivative of F is null.

For the purposes of this exposition we shall make an assumption which is in general fulfilled. We shall suppose that each critical point is *nondegenerate* in the sense that the terms F_2 of the second order in the Taylor's formula for F about the critical point is a nondegenerate quadratic form. Then, as in the elementary theory of conic sections, it is possible to make a real nonsingular linear transformation from the variables (u) to the variables (v) such that F_2 takes the form

$$F_2 = -v_1^2 - \cdots - v_k^2 + v_{k+1}^2 + \cdots + v_n^2.$$

The number k is called the *index* of the critical point.

A manifold such as Σ possesses an ith Betti number R_i ($i = 1, \cdots, n$). This is the maximum number of independent nonbounding i-cycles† on Σ. For example, if Σ is a torus then $R_0 = 1$, $R_1 = 2, R_2 = 1$. We shall be concerned with a 3-dimensional torus T_3. Such a manifold can be obtained by starting with a 2-dimensional torus T_2 and a 2-plane π_2 lying in a euclidean 3-plane, with π_2 not intersecting T_2. To obtain T_3 we revolve T_2 about π_2 in a 4-plane containing our 3-plane. Such a T_3 is sometimes called a product of three circles. For T_3 one has $R_0 = 1$, $R_1 = 3$, $R_2 = 3$,

---

† For details see Seifert-Threlfall, *Lehrbuch der Topologie*. Leipzig: 1934, Chap. III.

$R_3 = 1$. These numbers are the binomial coefficients when $n = 3$. We can obtain a 1–1 continuous image of an ordinary torus by identifying opposite sides of a square. Similarly one can obtain a 1–1 continuous image of T_3 by identifying opposite faces of a cube. With this identification, three mutually perpendicular edges of the cube represent three independent nonbounding 1-cycles, as can be shown. Similarly, three mutually perpendicular faces of the cube represent three independent nonbounding 2-cycles. A point is a 0-cycle and T_3 itself is a 3-cycle. In this way one intuitively accounts for the fact that $R_0 = 1$, $R_1 = 3$, $R_2 = 3$, $R_3 = 1$. A 3-dimensional manifold which is a 1–1 continuous image of T_3 will be called a *topological* 3-torus.

The theorem which will be used in what follows is that on Σ the number M_i of critical points of f of index i satisfies the fundamental relation†

$$(5.1) \qquad\qquad M_i \geqq R_i.$$

Thus on a topological 3-torus one can infer the existence of at least $1 + 3 + 3 + 1 = 8$ critical points.

Examples

1. *Triangles of light.* Let there be given three nonintersecting, simple, closed, nonsingular, analytic curves C_1, C_2, C_3 all lying in a 2-plane. We shall be concerned with triangles with vertices p_1, p_2, p_3 on C_1, C_2, C_3, respectively. Such a triangle will be called a *triangle of light* if a ray of light following this triangle is reflected at p_i as if C_i were a mirror, or if the angle in the triangle at p_i is π. How many triangles of light can we affirm to exist?

Let f be the sum of the lengths of the sides of the triangle $p_1 p_2 p_3$. We can refer C_i to a parameter u_i which is proportional to the arc length and varies from 0 to 2π. Then f becomes a function $f(u_1, u_2, u_3)$. The domain of definition of f is clearly a topological 3-torus. As a matter of analysis in the small, one proves by ele-

---

† See Morse, "Calculus of variations in the large." Colloquium lectures, *American Mathematical Society* (1934), Chap. VI. Also, Seifert-Threlfall, *Variationsrechnung im Grossen*. Leipzig: 1938.

mentary methods that f has a critical point if and only if the corresponding triangle is a triangle of light.

These triangles of light can then be classified according to the index of the corresponding critical point. According to relation (5.1) in the general theory of critical points there are at least $8 = 1 + 3 + 3 + 1$ of these triangles of light.

2. *Normals from a point to a topological 3-torus.* Let Σ_3 be a topological 3-torus in a euclidean 4-space. Let p be a fixed point not on Σ_3. We seek normals from p to Σ_3. To obtain these we let f be the distance from p to Σ_3 regarding f as a function of the point (u) of Σ_3. It can be shown that except for a subset of special points p the critical points of f are nondegenerate. Moreover, one then shows by a local analysis that f has a critical point (u) if and only if the line segment from p to (u) is normal to Σ_3 at (u). According to our general theorem there are then at least eight normals from p to Σ_3. These normals can be classified and it can be shown that the index of a nondegenerate critical point is the number of centers of principal curvature of Σ_3 between p and (u) on the given normal. Similar theorems hold for a topological 2-torus. Here the number of normals is at least 4.

3. *Three-planes passing through a fixed 2-plane and tangent to the preceding topological 3-torus* Σ_3. We suppose that the fixed 2-plane π_2 does not pass through a hole in Σ_3, that is, we suppose that π_2 can be moved indefinitely away from Σ_3 without intersecting Σ_3. We can then show that if π_2 is nonspecialized there are at least eight 3-planes through π_2 tangent to Σ_3.

4. *Heavy chain in equilibrium, with ends free to move on a topological 2-torus and on a closed curve C, respectively.* We suppose that the curve C and the topological 2-torus Σ_2 lie in euclidean 3-space but that no point of C and Σ_2 lie on the same plumb line. We suppose that a chain is provided which is larger than the maximum distance from a point of C to a point of Σ_2. The end points of the chain are supposed free to move on Σ_2 and C respectively and the chain is permitted to pass through Σ_2 or C. If the position of C is nonspecialized relative to Σ_2, then there are at least eight positions of equilibrium of the chain, seven of which are unstable. The function the critical points of which are sought gives the height of the center of gravity of the chain as a function of the end points of the chain.

The examples of this section are unified by the fact that the function involved is defined in each case on a topological 3-torus. The examples belong equally well to mechanics, geometry, or the calculus of variations. The restrictions as to nonspecialized positions of the configurations involved can all be removed by replacing the definition of a critical point in terms of derivatives by a topological definition of a critical point, and by replacing the classification of critical points according to their indices by a topological classification of a group theoretic character. This type of generalization both in its form and genesis is characteristic of analysis in the large.

It is possible that analysis in the large may eventually reduce to topology, but not until topology has been greatly broadened. It is equally conceivable that the apparently less general situations which arise with such frequency in problems in analysis in the large may form the canonical cases about which the topology of the future can be built.

Analysis is full of difficult but significant unsolved problems in the large. We mention only one example. How does the topological structure of the contour manifolds of the Jacobi least action integral J in the problem of three or more celestial bodies vary with the value of J? The independent variable in J is a closed path. The solution of this problem may disclose that the planetary orbits exist for essentially topological reasons. On the purely topological side, the number of problems the solution of which is necessary for a rapid advance of analysis in the large is very great, presenting a field that is virtually untouched.

REFERENCES

The reader wishing to enter this field should read the current introductions to topology and differential geometry. In the latter field, Reference 1 is recommended. Reference 2 introduces the reader to critical point theory and may well be followed by Reference 3. A description of the three different levels at which critical point theory proceeds is found in the fourth reference and others. The advanced reader can now turn to the fifth reference with its Section 7, "Application au calcul des variations." In Reference 6, the Lefschetz theorem is proved with the aid of the Morse theory of normals and focal points and the relation (5.1) in its appropriate form. Reference 7 contains one of the deepest applications of the critical point theory. In Reference 8, the analysis in the large of functions on a 3-torus, as applied in Examples 1 through 4, finds another kindred application.

1. Georges de Rham, *Variétés différentiables*. Paris: Hermann et Cie., 1955.

2. M. Morse and G. B. van Schaack, "The critical point theory under general boundary conditions," *Annals of Mathematics*, 35 (1934).

3. H. Seifert and W. Threlfall, *Variationsrechnung im Grossen*. Leipzig und Berlin: Teubner, 1938.

4. M. Morse, "Recent advances in variational theory in the large," *Proceedings of the International Congress of Mathematics*, (1950), pp. 143–55.

5. J.-P. Serre, "Homologie singulière des espaces fibrés," *Annals of Mathematics*, 54 (1951), pp. 425–505.

6. A. Andreotti and T. Frankel, "The Lefschetz theorem on hyperplane sections," *Annals of Mathematics*, 69 (1959), pp. 713–17.

7. R. Bott and H. Samelson, "Applications of the theory of Morse to symmetric spaces," *American Journal of Mathematics*, 70 (1958), pp. 964–1029.

8. Leon van Hove, "The occurrence of singularities in the elastic frequency distribution of a crystal," *Physical Review*, 89 (1953), pp. 1189–93.

9. John Milnor, "Morse theory" (based on lecture notes by M. Spivak and R. Wells), *Annals of Mathematics Series*, No. 51. Princeton, N.J.: Princeton University Press, 1963.

10. *Differential and Combinatorial Topology, A symposium in honor of Marston Morse*. Princeton Mathematical Series, Princeton, N.J.: Princeton University Press, 1965.

CURVES AND SURFACES IN EUCLIDEAN SPACE

S. S. Chern

1. INTRODUCTION

This article contains a treatment of some of the most elementary theorems in differential geometry in the large. They are the seeds for further developments and the subject should have a promising future. We shall consider the simplest cases, where the geometrical ideas are most clear.

1. THEOREM OF TURNING TANGENTS

Let E be the euclidean plane, which is oriented so that there is a prescribed sense of rotation. We define a smooth curve by expressing its position vector $X = (x_1, x_2)$ as a function of its arc length s. We suppose the function $X(s)$—that is, the functions $x_1(s)$, $x_2(s)$—to be twice continuously differentiable and the vector $X'(s)$ to be nowhere 0. The latter allows the definition of the unit tangent vector $e_1(s)$, which is the unit vector in the direction of $X'(s)$ and, since E is oriented, the unit normal vector $e_2(s)$, so that the rotation from e_1 to e_2 is positive. The vectors $X(s)$, $e_1(s)$, $e_2(s)$ are related by the so-called Frenet formulas

$$(1) \qquad \frac{dX}{ds} = e_1, \qquad \frac{de_1}{ds} = ke_2, \qquad \frac{de_2}{ds} = -ke_1.$$

The function $k(s)$ is called the *curvature*. It is defined together with its sign and changes its sign if the orientation of the curve or of the plane is reversed.

The curve C is called *closed*, if $X(s)$ is periodic of period L, L being the length of C. It is called *simple* if $X(s_1) \neq X(s_2)$, when $0 < s_1 - s_2 < L$. It is said to be *convex* if it lies in one side of every tangent line.

Let C be an oriented closed curve of length L, with the position vector $X(s)$ as a function of the arc length s. Let O be a fixed point in the plane, which we take as the origin of our coordinate system. Denote by Γ the unit circle about O. We define the tangential mapping $T : C \to \Gamma$ as the one which maps a point P of C to the endpoint of the unit vector through O parallel to the tangent vector to C at P. Obviously T is a continuous mapping. It is intuitively clear that when a point goes around C once its image point goes around Γ a number of times. This number will be called the rotation index of C. The theorem of turning tangents asserts that if C is simple, the rotation index is ± 1. We begin by giving a rigorous definition of the rotation index.

We choose a fixed vector through O, say Ox, and denote by $\tau(s)$ the angle which Ox makes with the vector $e_1(s)$. We assume that $0 \leqq \tau(s) < 2\pi$, so that $\tau(s)$ is uniquely determined. This function $\tau(s)$ is, however, not continuous, for in every neighborhood of s_0 at which $\tau(s_0) = 0$ there may be values of $\tau(s)$ differing from 2π by an arbitrarily small quantity. There exists nevertheless a continuous function $\bar{\tau}(s)$ closely related to $\tau(s)$, as given by the following lemma.

LEMMA: *There exists a continuous function $\bar{\tau}(s)$ such that $\bar{\tau}(s) \equiv \tau(s)$, mod 2π.*

Proof: To prove the lemma, we observe that the mapping T, being continuous, is uniformly continuous. Therefore, there exists a number $\delta > 0$, such that, for $|s_1 - s_2| < \delta$, $T(s_1)$ and $T(s_2)$ lie in the same open half-plane. From our conditions on $\bar{\tau}(s)$, it follows that, if $\bar{\tau}(s_1)$ is known, $\bar{\tau}(s_2)$ is completely determined. We divide the interval $0 \leqq s \leqq L$ by the points $s_0 (= 0) < s_1 < \cdots < s_m (= L)$ such that $|s_i - s_{i-1}| < \delta$, $i = 1, \cdots, m$. To define $\bar{\tau}(s)$, we assign to $\bar{\tau}(s_0)$ the value $\tau(s_0)$. Then it is determined in the subinterval $s_0 \leqq s \leqq s_1$, in particular at s_1, which determines it in the second subinterval, etc. The function $\bar{\tau}(s)$ so defined clearly satisfies the conditions of the lemma.

The difference $\bar{\tau}(L) - \bar{\tau}(0)$ is an integral multiple of 2π, say, $= \gamma 2\pi$. We assert that the integer γ is independent of the choice of the function $\bar{\tau}(s)$. In fact, let $\bar{\tau}'(s)$ be a function satisfying the same conditions. Then we have

$$\bar{\tau}'(s) - \bar{\tau}(s) = n(s) \cdot 2\pi,$$

where $n(s)$ is an integer. Since $n(s)$ is continuous in s, it must be a constant. It follows that

$$\bar{\tau}'(L) - \bar{\tau}'(0) = \bar{\tau}(L) - \bar{\tau}(0),$$

which proves the independence of γ from the choice of $\bar{\tau}(s)$. We define γ to be the rotation index of C. The *theorem of turning tangents* follows.

THEOREM: *The rotation index of a simple closed curve is* ± 1.

Proof: To prove this theorem, we consider the mapping Σ which carries an ordered pair of points of C, $X(s_1)$, $X(s_2)$, $0 \leqq s_1 \leqq s_2 \leqq L$, into the endpoint of the unit vector through O parallel to the secant joining $X(s_1)$ to $X(s_2)$. These ordered pairs of points can be represented as a triangle Δ in the (s_1, s_2)-plane defined by $0 \leqq s_1 \leqq s_2 \leqq L$. The mapping Σ of Δ into Γ is continuous. We also observe that its restriction to the side $s_1 = s_2$ is the tangential mapping T.

To a point $p \in \Delta$, let $\tau(p)$ be the angle which Ox makes with $O\Sigma(p)$, such that $0 \leqq \tau(p) < 2\pi$. Again this function need not be continuous. We shall, however, prove that there exists a continuous function $\bar{\tau}(p)$, $p \in \Delta$, such that $\bar{\tau}(p) \equiv \tau(p) \bmod 2\pi$.

In fact, let m be an interior point of Δ. We cover Δ by the radii through m. By the arguments used in the proof of the preceding lemma, we can define a function $\bar{\tau}(p)$, $p \in \Delta$, such that $\bar{\tau}(p) \equiv \tau(p)$, mod 2π, and such that it is continuous along every radius through m. It remains to prove that it is continuous in Δ. For this purpose, let p_0 be a point of Δ. Since Σ is continuous, it follows from the compactness of the segment mp_0 that there exists a number $\eta = \eta(p_0) > 0$, such that, for $q_0 \in mp_0$, and for any point of $q \in \Delta$ for which the distance $d(q, q_0) < \eta$, the points $\Sigma(q)$ and $\Sigma(q_0)$ are never antipodal. The latter condition is equivalent to the relation

$$(2) \qquad \bar{\tau}(q) - \bar{\tau}(q_0) \not\equiv 0, \quad \mathrm{mod}\ \pi.$$

Now let $\epsilon > 0$, $\epsilon < \pi/2$, be given. We choose a neighborhood U of p_0, such that U is contained in the η-neighborhood of p_0, and such that, for $p \in U$, the angle between $O\Sigma(p_0)$ and $O\Sigma(p)$ is less than ϵ. This is possible, because the mapping Σ is continuous. The last condition can be expressed in the form

$$(3) \qquad \bar{\tau}(p) - \bar{\tau}(p_0) = \epsilon' + 2k(p)\pi, \quad |\epsilon'| < \epsilon,$$

where $k(p)$ is an integer. Let q_0 be any point on the segment mp_0. Draw the segment q_0q parallel to p_0p, with q on mp. The function $\bar{\tau}(q) - \bar{\tau}(q_0)$ is continuous in q along mp and is zero when q coincides with m. Since $d(q, q_0)$ is less than η, it follows from Equation (2) that $|\bar{\tau}(q) - \bar{\tau}(q_0)| < \pi$. In particular, for $q_0 = p_0$, $|\bar{\tau}(p) - \bar{\tau}(p_0)| < \pi$. Combining this result with Equation (3), we get $k(p) = 0$, which proves that $\bar{\tau}(p)$ is continuous in Δ. Since $\bar{\tau}(p) \equiv \tau(p)$, mod 2π, it is easy to see that $\bar{\tau}(p)$ is differentiable.

Now let $A(0, 0)$, $B(0, L)$, and $D(L, L)$ be the vertices of Δ. The rotation index γ of C is defined by the line integral

$$2\pi\gamma = \int_{AD} d\bar{\tau}.$$

Since $\bar{\tau}(p)$ is defined in Δ, we have

$$\int_{AD} d\bar{\tau} = \int_{AB} d\bar{\tau} + \int_{BD} d\bar{\tau}.$$

To evaluate the line integrals on the right-hand side, we make use of a suitable coordinate system. We can suppose $X(0)$ to be the "lowest" point of C—that is, the point when the vertical coordinate is a minimum, and we choose $X(0)$ to be the origin O. The tangent vector to C at $X(0)$ is horizontal, and we call it Ox. The curve C then lies in the upper half-plane bounded by Ox, and the line integral $\int_{AB} d\bar{\tau}$ is equal to the angle rotated by OP as P traverses once along C. Since OP never points downward, this angle is $\epsilon\pi$, with $\epsilon = \pm 1$. Similarly, the integral $\int_{BD} d\bar{\tau}$ is the angle rotated by PO as P goes once along C. Its value is also equal to $\epsilon\pi$. Hence, the sum of the two integrals is $\epsilon 2\pi$ and the rotation index of C is ± 1, which completes our proof.

We can also define the rotation index by an integral formula. In fact, using the function $\bar{\tau}(s)$ in our lemma, we can express the components of the unit tangent and normal vectors as follows:

$$e_1 = (\cos\bar{\tau}(s), \sin\bar{\tau}(s)), \qquad e_2 = (-\sin\bar{\tau}(s), \cos\bar{\tau}(s)).$$

It follows that

$$d\bar{\tau}(s) = de_1 \cdot e_2 = k\, ds.$$

From this equation, we derive the following formula for the rotation index:

$$(4) \qquad\qquad 2\pi\gamma = \int_C k\, ds.$$

This formula holds for closed curves which are not necessarily simple.

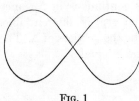

FIG. 1

The accompanying figure gives an example of a closed curve with rotation index zero.

Many interesting theorems in differential geometry are valid for a more general class of curves, the so-called *sectionally smooth curves*. Such a curve is the union of a finite number of smooth arcs A_0A_1, A_1A_2, $\cdots$, $A_{m-1}A_m$, where the tangents of the two arcs through a common vertex A_i, $i = 1, \cdots, m - 1$, may be different. The curve is called *closed*, if $A_0 = A_m$. The simplest example of a closed sectionally smooth curve is a rectilinear polygon.

The notion of rotation index and the theorem of turning tangents can be extended to closed sectionally smooth curves; we summarize, without proof, the result as follows. Let s_i, $i = 1, \cdots, m$, be the arc length measured from A_0 to A_i, so that $s_m = L$ is the length of the curve. The curve supposedly being oriented, the tangential mapping is defined at all points different from A_i. At a vertex A_i there are two unit vectors, tangent respectively to $A_{i-1}A_i$ and A_iA_{i+1}. (We define $A_{m+1} = A_1$.) The corresponding points on Γ we denote by $T(A_i)^-$ and $T(A_i)^+$. Let φ_i be the angle from $T(A_i)^-$ to $T(A_i)^+$, with $0 < \varphi_i < \pi$, briefly the exterior angle from the tangent to $A_{i-1}A_i$ to the tangent to A_iA_{i+1}. For each arc $A_{i-1}A_i$, a continuous function $\tilde{\tau}(s)$ can be defined which is one of the determinations of the angle from Ox to the tangent at $X(s)$. The number γ defined by the equation

$$(5) \qquad 2\pi\gamma = \sum_{i=1}^{m} \{\tilde{\tau}(s_i) - \tilde{\tau}(s_{i-1})\} + \sum_{i=1}^{m} \varphi_i$$

is an integer, which will be called the *rotation index* of the curve. The theorem of turning tangents is again valid.

THEOREM. *If a sectionally smooth curve is simple, the rotation index is equal to* ± 1.

As an application of the theorem of turning tangents, we wish to give the following characterization of a simple closed convex curve.

REMARK: *A simple closed curve is convex, if and only if it can be so oriented that its curvature is greater than, or equal to, 0.*

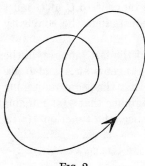

Let us first remark that the theorem is not true without the assumption that the curve is simple. In fact, the accompanying figure gives a non-convex curve with $k > 0$.

Proof: To prove the theorem, we let $\tilde{\tau}(s)$ be the function constructed, so that we have $k = d\tilde{\tau}/ds$. The condition $k \geqq 0$ is equivalent to the assertion that $\tilde{\tau}(s)$ is a monotone non-decreasing function. Because C is

FIG. 2

simple, we can suppose that $\bar{\tau}(s)$, $0 \leqq s \leqq L$, increases from 0 to 2π. It follows that if the tangents at $X(s_1)$ and $X(s_2)$, $0 \leqq s_1 < s_2 < L$, are parallel in the same sense, the arc of C from $X(s_1)$ to $X(s_2)$ is a straight line segment and these tangents must coincide.

Suppose $\bar{\tau}(s)$, $0 \leqq s \leqq L$, is monotone nondecreasing and C is not convex. There is a point $A = X(s_0)$ on C such that there are points of C at both sides of the tangent t to C at A. Choose a positive side of t and consider the oriented perpendicular distance from a point $X(s)$ of C to t. This is a continuous function in s and attains a maximum and a minimum at the points M and N of C, respectively. Clearly M and N are not on t and the tangents to C at M and N are parallel to t. Among these two tangents and t itself, there are two tangents parallel in the same sense, which, according to the preceding remark, is impossible.

Next we let C be convex. To prove that $\bar{\tau}(s)$ is monotone, we suppose $\bar{\tau}(s_1) = \bar{\tau}(s_2)$, $s_1 < s_2$. Then the tangents at $X(s_1)$ and $X(s_2)$ are parallel in the same sense. But there exists a tangent parallel to them in the opposite sense. From the convexity of C it follows that two of them coincide.

We are thus led to the consideration of a line t tangent to C at two distinct points, A and B. We claim that the segment AB must be a part of C. In fact, suppose this is not the case and let D be a point of AB not on C. Draw through D a perpendicular u to t in the half-plane which contains C. Then u intersects C in at least two points. Among these points of intersection, let F be the farthest from t and G the nearest, so that $F \neq G$. Then G is an interior point of the triangle ABF. The tangent to C at G must have points of C in both sides, which contradicts the convexity of C.

It follows that, under the hypothesis of the last paragraph, the segment AB is a part of C and that the tangents at A and B are parallel in the same sense. This proves that the segment joining $X(s_1)$ to $X(s_2)$ belongs to C. The latter implies that $\bar{\tau}(s)$ remains constant in the interval $s_1 \leqq s \leqq s_2$. Hence, the function $\bar{\tau}(s)$ is monotone, and our theorem is proved.

The first half of the theorem can also be stated as follows.

REMARK: *A closed curve with $k(s) \geqq 0$ and rotation index equal to 1 is convex.*

The theorem of turning tangents was essentially known to Riemann. The above proof was given by H. Hopf, *Compositio Mathematica* 2 (1935), pp. 50–62. For further reading, see:

1. H. Whitney, "On regular closed curves in the plane," *Compositio Mathematica* 4 (1937), pp. 276–84.

2. S. Smale, "Regular curves on a Riemannian manifold," *Transactions of the American Mathematical Society* 87 (1958), pp. 492–511.

3. S. Smale, "A classification of immersions of the two-sphere," *Transactions of the American Mathematical Society* 90 (1959), pp. 281–90.

2. THE FOUR-VERTEX THEOREM

An interesting theorem on closed plane curves is the so-called "four-vertex theorem." By a *vertex* of an oriented closed plane curve we mean a point at which the curvature has a relative extremum. Since the curve forms a compact point set, it has at least two vertices, corresponding respectively to the absolute minimum and maximum of the curvature. Our theorem says that there are at least four.

THEOREM: *A simple closed convex curve has at least four vertices.*

This theorem was first presented by Mukhopadhyaya (1909); the proof we shall give was the work of G. Herglotz. It is also true for nonconvex curves, but the proof is more difficult. The theorem cannot be improved, because an ellipse with unequal axes has exactly four vertices, which are its points of intersection with the axes.

Proof: We suppose that the curve C has only two vertices, M and N, and we shall show that this leads to a contradiction. The line MN does not meet C in any other point, for if it does, the tangent line to C at the middle point must contain the other two points. By the last section, this condition is possible only when the segment MN is a part of C. It would follow that the curvature vanishes at M and N, which is not possible, since they are the points where the curvature takes the absolute maximum and minimum respectively.

We denote by 0 and s_0 the parameters of M and N respectively and take MN to be the x_1-axis. Then we can suppose

$$x_2(s) < 0, \qquad 0 < s < s_0,$$
$$x_2(s) > 0, \qquad s_0 < s < L,$$

where L is the length of C. Let $(x_1(s), x_2(s))$ be the position vector of a point of C with the parameter s. Then the unit tangent and normal vectors have the components

$$e_1 = (x_1', x_2'), \qquad e_2 = (-x_2', x_1'),$$

where primes denote differentiations with respect to s. From the Frenet formulas we get

(6) $$x_1'' = -kx_2', \qquad x_2'' = kx_1'.$$

It follows that

$$\int_0^L kx_2' \, ds = -x_1' \Big|_0^L = 0.$$

The integral in the left-hand side can be written as a sum:

$$\int_0^L kx_2' \, ds = \int_0^{s_0} kx_2' \, ds + \int_{s_0}^L kx_2' \, ds.$$

To each summand we apply the second mean value theorem, which is stated as follows. Let $f(x)$, $g(x)$, $a \leqq x \leqq b$, be two functions in x such that $f(x)$ and $g'(x)$ are continuous and $g(x)$ is monotone. Then there exists ξ, $a < \xi < b$, satisfying the equation,

$$\int_a^b f(x)g(x) \, dx = g(a) \int_a^\xi f(x) \, dx + g(b) \int_\xi^b f(x) \, dx.$$

Since $k(s)$ is monotone in each of the intervals $0 \leqq s \leqq s_0$, $s_0 \leqq s \leqq L$, we get

$$\int_0^{s_0} kx_2' \, ds = k(0) \int_0^{\xi_1} x_2' \, ds + k(s_0) \int_{\xi_1}^{s_0} x_2' \, ds$$
$$= x_2(\xi_1)(k(0) - k(s_0)), \qquad 0 < \xi_1 < s_0$$

$$\int_{s_0}^{L} kx_2' \, ds = k(s_0) \int_{s_0}^{\xi_2} x_2' \, ds + k(L) \int_{\xi_2}^{L} x_2' \, ds$$

$$= x_2(\xi_2)(k(s_0) - k(0)), \qquad s_0 < \xi_2 < L.$$

Since the sum of the left-hand members is zero, these equations give

$$(x_2(\xi_1) - x_2(\xi_2))(k(0) - k(s_0)) = 0,$$

which is a contradiction, because

$$x_2(\xi_1) - x_2(\xi_2) < 0, \qquad k(0) - k(s_0) > 0.$$

It follows that there is at least one more vertex on C. Since the relative extrema occur in pairs, there are at least four vertices and the theorem is proved.

At a vertex we have $k' = 0$. Hence, we can also say that on a simple closed convex curve there are at least four points at which $k' = 0$.

The four-vertex theorem is also true for simple closed nonconvex plane curves; see:

1. S. B. Jackson, "Vertices for plane curves," *Bulletin of the American Mathematical Society* 50 (1944), pp. 564–578.

2. L. Vietoris, "Ein einfacher Beweis des Vierscheitelsatzes der ebenen Kurven," *Archiv der Mathematik* 3 (1952), pp. 304–306.

For further reading, see:

1. P. Scherk, "The four-vertex theorem," *Proceedings of the First Canadian Mathematical Congress.* Montreal: 1945, pp. 97–102.

3. ISOPERIMETRIC INEQUALITY
FOR PLANE CURVES

The theorem can be stated as follows.

THEOREM: *Among all simple closed curves having a given length the circle bounds the largest area. In other words, if L is the length of a simple closed curve C, and A is the area it bounds, then*

(7) $$L^2 - 4\pi A \geqq 0.$$

Moreover, the equality sign holds only when C is a circle.

Many proofs have been given of this theorem, differing in degree of elegance and in the range of curves under consideration—that

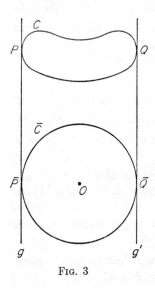

FIG. 3

is, whether differentiability or convexity is supposed. We shall give two proofs, the work of E. Schmidt (1939) and A. Hurwitz (1902), respectively.

Schmidt's Proof: We enclose C between two parallel lines, g and g', such that C lies between g and g' and is tangent to them at the points P and Q, respectively. We let $s = 0$, s_0 being the parameters of P and Q, and construct a circle $\overline{C}$ tangent to g and g' at $\overline{P}$ and $\overline{Q}$, respectively. Denote its radius by r and take its center to be the origin of a coordinate system. Let $X(s) = (x_1(s), x_2(s))$ be the position vector of C, so that $(x_1(0), x_2(0)) = (x_1(L), x_2(L))$. As the position vector of $\overline{C}$ we take $(\overline{x}_1(s), x_2(s))$, such that

$$
\begin{aligned}
\overline{x}_1(s) &= x_1(s), \\
(8) \qquad \overline{x}_2(s) &= -\sqrt{r^2 - x_1^2(s)}, \quad 0 \leqq s \leqq s_0 \\
&= +\sqrt{r^2 - x_1^2(s)}, \quad s_0 \leqq s \leqq L.
\end{aligned}
$$

Denote by $\overline{A}$ the area bounded by $\overline{C}$. Now the area bounded by a closed curve can be expressed by the line integral

$$
A = \int_0^L x_1 x_2' \, ds = -\int_0^L x_2 x_1' \, ds = \tfrac{1}{2} \int_0^L (x_1 x_2' - x_2 x_1') \, ds.
$$

Applying this to our two curves C and $\overline{C}$, we get

$$
A = \int_0^L x_1 x_2' \, ds
$$

$$
\overline{A} = \pi r^2 = -\int_0^L \overline{x}_2 \overline{x}_1' \, ds = -\int_0^L \overline{x}_2 x_1' \, ds.
$$

Adding these two equations, we have

$$A + \pi r^2 = \int_0^L (x_1 x_2' - \overline{x}_2 x_1') \, ds \leqq \int_0^L \sqrt{(x_1 x_2' - \overline{x}_2 x_1')^2} \, ds$$

$$(9) \qquad \leqq \int_0^L \sqrt{(x_1^2 + \overline{x}_2^2)(x_1'^2 + x_2'^2)} \, ds$$

$$= \int_0^L \sqrt{x_1^2 + \overline{x}_2^2} \, ds = Lr.$$

Since the geometric mean of two positive numbers is less than or equal to their arithmetic mean, it follows that

$$\sqrt{A} \sqrt{\pi r^2} \leqq \tfrac{1}{2}(A + \pi r^2) \leqq \tfrac{1}{2} Lr,$$

which gives, after squaring and cancellation of r^2, the inequality in Equation (7).

Suppose now that the equality sign in Equation (7) holds; then A and πr^2 have the same geometric and arithmetic mean, so that $A = \pi r^2$ and $L = 2\pi r$. The direction of the lines g and g' being arbitrary, this means that C has the same "width" in all directions. Moreover, we must have the equality sign everywhere in Equation (9). It follows, in particular, that

$$(x_1 x_2' - \overline{x}_2 x_1')^2 = (x_1^2 + \overline{x}_2^2)(x_1'^2 + x_2'^2),$$

which gives

$$\frac{x_1}{x_2'} = \frac{-\overline{x}_2}{x_1'} = \frac{\sqrt{x_1^2 + \overline{x}_2^2}}{\sqrt{x_1'^2 + x_2'^2}} = \pm r.$$

From the first equality in Equation (9), the factor of proportionality is seen to be r, that is,

$$x_1 = r x_2', \qquad \overline{x}_2 = -r x_1',$$

which remains true when we interchange x_1 and x_2, so that

$$x_2 = r x_1'.$$

Therefore, we have

$$x_1^2 + x_2^2 = r^2,$$

which means that C is a circle.

Hurwitz's proof makes use of the theory of Fourier series. We shall first prove the lemma of Wirtinger.

LEMMA: *Let $f(t)$ be a continuous periodic function of period 2π, possessing a continuous derivative $f'(t)$. If $\int_0^{2\pi} f(t)\, dt = 0$, then*

$$(10) \qquad\qquad \int_0^{2\pi} f'(t)^2\, dt \geqq \int_0^{2\pi} f(t)^2\, dt.$$

Moreover, the equality sign holds if and only if

$$(11) \qquad\qquad f(t) = a \cos t + b \sin t.$$

Proof: To prove the lemma, we let the Fourier series expansion of $f(t)$ be

$$f(t) \sim \frac{a_0}{2} + \sum_{n=1}^{\infty} (a_n \cos nt + b_n \sin nt).$$

Since $f'(t)$ is continuous, its Fourier series can be obtained by differentiation term by term, and we have

$$f'(t) \sim \sum_{n=1}^{\infty} (nb_n \cos nt - na_n \sin nt).$$

Since

$$\int_0^{2\pi} f(t)\, dt = \pi a_0,$$

it follows from our hypothesis that $a_0 = 0$. By Parseval's formula, we get

$$\int_0^{2\pi} f(t)^2\, dt = \sum_{n=1}^{\infty} (a_n^2 + b_n^2),$$

$$\int_0^{2\pi} f'(t)^2\, dt = \sum_{n=1}^{\infty} n^2(a_n^2 + b_n^2).$$

Hence,

$$\int_0^{2\pi} f'(t)^2\, dt - \int_0^{2\pi} f(t)^2\, dt = \sum_{n=1}^{\infty} (n^2 - 1)(a_n^2 + b_n^2),$$

which is greater than, or equal to, 0. It is equal to zero, only if $a_n = b_n = 0$ for all $n > 1$. Therefore, $f(t) = a_1 \cos t + b_1 \sin t$, which proves the lemma.

Hurwitz's Proof: In order to prove the inequality in Equation (7), we assume, for simplicity, that $L = 2\pi$, and that

$$\int_0^{2\pi} x_1(s)\, ds = 0.$$

The latter means that the center of gravity lies on the x_1-axis, a condition which can always be achieved by a proper choice of the coordinate system. The length and the area are given by the integrals,

$$2\pi = \int_0^{2\pi} (x_1'^2 + x_2'^2) \, ds, \quad \text{and} \quad A = \int_0^{2\pi} x_1 x_2' \, ds.$$

From these two equations we get

$$2(\pi - A) = \int_0^{2\pi} (x_1'^2 - x_1^2) \, ds + \int_0^{2\pi} (x_1 - x_2')^2 \, ds.$$

The first integral is greater than, or equal to, 0 by our lemma and the second integral is clearly greater than, or equal to, 0. Hence, $A \leqq \pi$, which is our isoperimetric inequality.

The equality sign holds only when

$$x_1 = a \cos s + b \sin s, \qquad x_2' = x_1,$$

which gives

$$x_1 = a \cos s + b \sin s, \qquad x_2 = a \sin s - b \cos s + c.$$

Thus, C is a circle.

For further reading, see:

1. E. Schmidt, "Beweis der isoperimetrischen Eigenschaft der Kugel im hyperbolischen und sphärischen Raum jeder Dimensionenzahl," *Math. Zeit.* 49 (1943), pp. 1–109.

4. TOTAL CURVATURE OF A SPACE CURVE

The *total curvature* of a closed space curve C of length L is defined by the integral

$$(12) \qquad \mu = \int_0^L |k(s)| \, ds,$$

where $k(s)$ is the curvature. For a space curve, only $|k(s)|$ is defined.

Suppose C is oriented. Through the origin O of our space we draw vectors of length 1 parallel to the tangent vectors of C. Their end-points describe a closed curve Γ on the unit sphere, to be called the *tangent indicatrix* of C. A point of Γ is singular (that is, with either no tangent or a tangent of higher contact) if it is the image of a point of zero curvature of C. Clearly the total curvature of C is equal to the length of Γ.

Fenchel's theorem concerns the total curvature.

THEOREM: *The total curvature of a closed space curve C is greater than, or equal to, 2π. It is equal to 2π if and only if C is a plane convex curve.*

The following proof of this theorem was found independently by B. Segre (*Bolletino della Unione Matematica Italiana* 13 (1934), 279–283), and by H. Rutishauser and H. Samelson (*Comptes Rendus Hebdomadaires des Séances de l'Académie des Sciences* 227 (1948), 755–757). See also W. Fenchel, *Bulletin of the American Mathematical Society* 57 (1951), 44–54. The proof depends on the following lemma:

LEMMA: *Let Γ be a closed rectifiable curve on the unit sphere, with length $L < 2\pi$. There exists a point m on the sphere such that the spherical distance $\overline{mx} \leq L/4$ for all points x of Γ. If Γ is of length 2π but is not the union of two great semicircular arcs, there exists a point m such that $\overline{mx} < \pi/2$ for all x of Γ.*

We use the notation $\overline{ab}$ to denote the spherical distance of two points, a and b. If $\overline{ab} < \pi$, their *midpoint* m is the point defined by the conditions $\overline{am} = \overline{bm} = \frac{1}{2}\overline{ab}$. Let x be a point such that $\overline{mx} \leq \frac{1}{2}\pi$. Then $2\overline{mx} \leq \overline{ax} + \overline{bx}$. In fact, let x' be the symmetry of x relative to m. Then,

$$\overline{x'a} = \overline{xb}, \qquad \overline{x'x} = \overline{x'm} + \overline{mx} = 2\overline{mx}.$$

If we use the triangle inequality, it follows that

$$(13) \qquad 2\overline{mx} = \overline{x'x} \leq \overline{x'a} + \overline{ax} = \overline{ax} + \overline{bx},$$

as to be proved.

Lemma Proof: To prove the first part of the lemma, we take two points, a and b, on Γ which divide the curve into two equal arcs. Then $\overline{ab} < \pi$, and we denote the midpoint by m. Let x be a point of Γ such that $2\overline{mx} < \pi$. Such points exist—for example, the point a. Then we have

$$\overline{ax} \leqq \widehat{ax}, \qquad \overline{bx} \leqq \widehat{bx},$$

where $\widehat{ax}$ and $\widehat{bx}$ are respectively the arc lengths along Γ. From Equation (13), it follows that

$$2\overline{mx} \leqq \widehat{ax} + \widehat{bx} = \widehat{ab} = \frac{L}{2}.$$

Hence, the function $f(x) = \overline{mx}$, $x \in \Gamma$, is either $\geqq \pi/2$ or $\leqq L/4 < \pi/2$. Since Γ is connected and $f(x)$ is a continuous function in Γ, the range of the function $f(x)$ is connected in the interval $(0, \pi)$. Therefore, we have $f(x) = \overline{mx} \leqq L/4$.

Consider next the case that Γ is of length 2π. If Γ contains a pair of antipodal points, then, being of length 2π, it must be the union of two great semicircular arcs. Suppose that there is a pair of points, a and b, which bisect Γ such that

$$\overline{ax} \mid \overline{bx} < \pi$$

for all $x \in \Gamma$. Again, let m denote the midpoint of a and b. If $f(x) = \overline{mx} \leqq \frac{1}{2}\pi$, we have, from Equation (13),

$$2\overline{mx} \leqq \overline{ax} + \overline{bx} < \pi,$$

which means that $f(x)$ omits the value $\pi/2$. Since its range is connected and since $f(a) < \pi/2$, we have $f(x) < \pi/2$ for all $x \in \Gamma$. Thus the lemma is true in this case.

It remains to consider the case that Γ contains no pair of antipodal points, and that for any pair of points a and b which bisect Γ, there is a point $x \in \Gamma$ with

$$\overline{ax} + \overline{bx} = \pi.$$

An elementary geometrical argument, which we leave to the reader, will show that this is impossible. Thus, the lemma is proved.

Theorem Proof: To prove Fenchel's theorem, we take a fixed unit vector A and put

$$g(s) = AX(s),$$

where the right-hand side denotes the scalar product of the vectors A and $X(s)$. The function $g(s)$ is continuous on C and hence must have a maximum and a minimum. Since $g'(s)$ exists, we have, at such an extremum s_0,

$$g'(s_0) = AX'(s_0) = 0.$$

Thus A, as a point on the unit sphere, has a distance $\pi/2$ from at least two points of the tangent indicatrix. Since A is arbitrary, the tangent indicatrix is met by every great circle. It follows from the lemma that its length is greater than, or equal to, 2π.

Suppose next that the tangent indicatrix Γ is of length 2π. By our lemma, it must be the union of two great semicircular arcs. It follows that C itself is the union of two plane arcs. Since C has a tangent everywhere, it must be a plane curve. Suppose C be so oriented that its rotation index

$$\frac{1}{2\pi} \int_0^L k \, ds \geqq 0.$$

Then we have

$$0 \leqq \int_0^L \{|k| - k\} \, ds = 2\pi - \int_0^L k \, ds$$

so that the rotation index is either 0 or 1. To a given vector in the plane there is parallel to it a tangent t of C such that C lies to the left of t. Then t is parallel to the vector in the same sense, and at its point of contact we have $k \geqq 0$, implying that $\int_{k>0} k \, ds \geqq 2\pi$. Since $\int_C |k| \, ds = 2\pi$, there is no point with $k < 0$, and $\int k \, ds = 2\pi$. From the remark at the end of Section 1, we conclude that C is convex.

As a corollary we have the following theorem.

COROLLARY: *If $|k(s)| \leqq 1/R$ for a closed space curve C, C has a length $L \geqq 2\pi R$.*

We have

$$L = \int_0^L ds \geqq \int_0^L R|k| \, ds = R \int_0^L |k| \, ds \geqq 2\pi R.$$

Fenchel's theorem holds also for sectionally smooth curves. As the total curvature of such a curve we define

$$(14) \qquad \mu = \int_0^L |k| \, ds + \sum_i a_i$$

where the a_i are the angles at the vertices. In other words, in this case the tangent indicatrix consists of a number of arcs each corresponding to a smooth arc of C; we join successive vertices by the shortest great circular arc on the unit sphere. The length of the curve so obtained is the total curvature of C. It can be proved that for a closed sectionally smooth curve we have also $\mu \geqq 2\pi$.

We wish to give another proof of Fenchel's theorem and a related theorem of Fary-Milnor on the total curvature of a knot.[†] The basis is Crofton's theorem on the measure of great circles which cut an arc on the unit sphere. Every oriented great circle determines uniquely a "pole," the endpoint of the unit vector normal to the plane of the circle. By the measure of a set of great circles on the unit sphere is meant the area of the domain of their poles. Then Crofton's theorem is stated as follows.

THEOREM: *Let* Γ *be a smooth arc on the unit sphere* Σ_0. *The measure of the oriented great circles of* Σ_0 *which meet* Γ, *each counted a number of times equal to the number of its common points with* Γ, *is equal to four times the length of* Γ.

Proof: We suppose Γ is defined by a unit vector $e_1(s)$ expressed as a function of its arc length s. Locally (that is, in a certain neighborhood of s), let $e_2(s)$ and $e_3(s)$ be unit vectors depending smoothly on s, such that the scalar products

$$(15) \qquad e_i \cdot e_j = \delta_{ij}, \quad 1 \leqq i, j \leqq 3$$

and

$$(16) \qquad \det (e_1, e_2, e_3) = +1.$$

Then we have

$$(17) \qquad \begin{cases} \dfrac{de_1}{ds} = a_2e_2 + a_3e_3, \\[2mm] \dfrac{de_2}{ds} = -a_2e_1 + a_1e_3, \\[2mm] \dfrac{de_3}{ds} = -a_3e_1 - a_1e_2. \end{cases}$$

---

† I. Fary (*Bulletin de la Société Mathématique de France*, 77 (1949), pp. 128–138), and J. Milnor (*Annals of Mathematics*, 52 (1950), pp. 248–257).

The skew-symmetry of the matrix of the coefficients in the above system of equations follows from differentiation of Equations (15). Since s is the arc length of Γ, we have

$$(18) \qquad\qquad a_2^2 + a_3^2 = 1,$$

and we put

$$(19) \qquad\qquad a_2 = \cos \tau(s), \qquad a_3 = \sin \tau(s).$$

If an oriented great circle meets Γ at the point $e_1(s)$, its pole is of the form $Y = \cos \theta\, e_2(s) + \sin \theta \cdot e_3(s)$, and vice versa. Thus (s, θ) serve as local coordinates in the domain of these poles; we wish to find an expression for the element of area of this domain.

For this purpose, we write

$$dY = (-\sin \theta\, e_2 + \cos \theta\, e_3)(d\theta + a_1\, ds) - e_1(a_2 \cos \theta + a_3 \sin \theta)\, ds.$$

Since $-\sin \theta\, e_2 + \cos \theta\, e_3$ and e_1 are two unit vectors orthogonal to Y, the element of area of Y is

$$(20) \quad |dA| = |a_2 \cos \theta + a_3 \sin \theta|\, d\theta\, ds = |\cos (\tau - \theta)|\, d\theta\, ds,$$

where the absolute value at the left-hand side means that the area is calculated in the measure-theoretic sense, with no regard to orientation. To the point Y let $Y^\perp$ be the oriented great circle with Y as its pole, and let $n(Y^\perp)$ be the (arithmetic) number of points common to $Y^\perp$ and Γ. Then the measure μ in our theorem is given by

$$\mu = \int n(Y^\perp)|dA| = \int_0^\lambda ds \int_0^{2\pi} |\cos (\tau - \theta)|\, d\theta,$$

where λ is the length of Γ. As θ ranges from 0 to 2π, the variation of $|\cos (\tau - \theta)|$, for a fixed s, is 4. Hence, we get $\mu = 4\lambda$, which proves Crofton's theorem.

By applying the theorem to each subarc and adding, we see that the theorem remains true when Γ is a sectionally smooth curve on the unit sphere. Actually, the theorem is true for any rectifiable arc on the sphere, but the proof is much longer.

For a closed space curve the tangent indicatrix of which fulfills the conditions of Crofton's theorem, Fenchel's theorem is an easy consequence. In fact, the proof of Fenchel's theorem shows us that the tangent indicatrix of a closed space curve meets every great circle in at least two points—that is, $n(Y^\perp) \geqq 2$. It follows that its length is

$$\lambda = \int |k|\,ds = \tfrac{1}{4} \int n(Y^\perp)|dA| \geqq 2\pi,$$

because the total area of the unit sphere is 4π.

Crofton's theorem also leads to the following theorem of Fary and Milnor, which gives a necessary condition on the total curvature of a knot.

THEOREM: *The total curvature of a knot is greater than, or equal to, 4π.*

Since $n(Y^\perp)$ is the number of relative maxima or minima of the "height function," $Y \cdot X(s)$, it is even. Suppose that the total curvature of a closed space curve C is $< 4\pi$. There exists $Y \in \Sigma_0$, such that $n(Y^\perp) = 2$. By a rotation, suppose Y is the point $(0, 0, 1)$. Then the function $x_3(s)$ has only one maximum and one minimum. These points divide C into two arcs, such that x_3 increases on the one and decreases on the other. Every horizontal plane between the two extremal horizontal planes meets C in exactly two points. If we join them by a segment, all these segments will form a surface which is homeomorphic to a circular disk, which proves that C is not knotted.

For further reading, see:

1. S. S. Chern and R. K. Lashof, "On the total curvature of immersed manifolds," I, *American Journal of Mathematics* 79 (1957), pp. 302–18, and II, *Michigan Mathematical Journal* 5 (1958), pp. 5–12.

2. N. H. Kuiper, "Convex immersions of closed surfaces in E^5," *Comm. Math. Helv.* 35 (1961), pp. 85–92.

On integral geometry compare the article of Santalo in this volume.

5. DEFORMATION OF A SPACE CURVE

It is well-known that a one-one correspondence between two curves under which the arc lengths, the curvatures (when not equal to 0), and the torsions are respectively equal, can only be established by a proper motion. It is natural to study the correspondences under which only s and k are equal. We shall call such a correspondence a deformation of the space curve (in German, *Vorwindung*). The most notable result in this direction is a theorem of A. Schur, which formulates the geometrical fact that if an arc is "stretched," the distance between its endpoints becomes longer. Using the name curvature to mean here always its absolute value, we state Schur's theorem as follows.

THEOREM: *Let C be a plane arc with the curvature $k(s)$ which forms a convex curve with its chord, AB. Let C^* be an arc of the same length referred to the same parameter s such that its curvature $k^*(s) \leqq k(s)$. If d^* and d denote the lengths of the chords joining their endpoints, then $d \leqq d^*$. Moreover, the equality sign holds when and only when C and C^* are congruent.*

Proof: Let Γ and Γ^* be the tangent indicatrices of C and C^* respectively, P_1 and P_2 two points on Γ, and P_1^* and P_2^* their corresponding points on Γ^*. We denote by $\widehat{P_1P_2}$ and $\widehat{P_1^*P_2^*}$ their arc lengths and by $\overline{P_1P_2}$ and $\overline{P_1^*P_2^*}$ their spherical distances. Then we have

$$\overline{P_1P_2} \leqq \widehat{P_1P_2}, \qquad \overline{P_1^*P_2^*} \leqq \widehat{P_1^*P_2^*}.$$

The inequality on the curvature implies

$$(21) \qquad \widehat{P_1^*P_2^*} \leqq \widehat{P_1P_2}.$$

Since C is convex, Γ lies on a great circle, and we have

$$\overline{P_1P_2} = \widehat{P_1P_2},$$

provided that $\overline{P_1P_2} \leqq \pi$. Now let Q be a point on C at which the tangent is parallel to the chord. Denote by P_0 its image point on Γ. Then the condition $\overline{P_0P} \leqq \pi$ is satisfied by any point P on Γ, and if P_0^* denotes the point on Γ^* corresponding to P_0, we have

$$(22) \qquad \overline{P_0^*P^*} \leqq \overline{P_0P},$$

from which it follows that

$$(23) \qquad \cos \overline{P_0^*P^*} \geqq \cos \overline{P_0P},$$

since the cosine function is a monotone decreasing function of its argument when the latter lies between 0 and π.

Because C is convex, d is equal to the projection of C on its chord:

$$(24) \qquad d = \int_0^L \cos \overline{P_0 P} \, ds.$$

On the other hand, we have

$$(25) \qquad d^* \geq \int_0^L \cos \overline{P_0^* P^*} \, ds,$$

for the integral on the right-hand side is equal to the projection of C^*, and hence of the chord joining its endpoints, on the tangent at the point Q^* corresponding to Q. Combining Equations (23), (24), and (25), we get $d^* \geq d$.

Suppose that $d = d^*$. Then the inequalities in Equations (22), (23), and (25) become equalities, and the chord joining the endpoints A^* and B^* of C^* must be parallel to the tangent at Q^*. In particular, we have

$$\overline{P_0^* P^*} = \overline{P_0 P},$$

which implies that the arcs A^*Q^* and B^*Q^* are plane arcs. On the other hand, we have, by using Equation (21),

$$\overline{P_0^* P^*} \leq \widehat{P_0^* P^*} \leq \widehat{P_0 P} = \overline{P_0 P},$$

or

$$\widehat{P_0^* P^*} = \widehat{P_0 P}.$$

Hence, the arcs A^*Q^* and B^*Q^* have the same curvature as AQ and BQ at corresponding points and are therefore respectively congruent.

It remains to prove that the arcs A^*Q^* and B^*Q^* lie in the same plane. Suppose the contrary. They must be tangent at Q^* to the line of intersection of the two distinct planes on which they lie. Since this line is parallel to A^*B^*, the only possibility is that it contains A^* and B^*; however, then the tangent to C at Q must also contain the endpoints A and B, which is a contradiction. Hence, C^* is a plane arc and is congruent to C.

Schur's theorem has many applications. For example, it gives a solution of the following minimum problem: Determine the shortest closed curve with a curvature $k(s) \leqq 1/R$, R being a constant. The answer is, of course, a circle.

REMARK: *The shortest closed curve with curvature $k(s) \leqq 1/R$, R being a constant, is a circle of radius R.*

By the corollary to Fenchel's theorem, such a curve has length $2\pi R$. Comparing it with a circle of radius R, we conclude from Schur's theorem (with $d^* = d = 0$) that it must itself be a circle.

As a second application of Schur's theorem, we shall derive a theorem of Schwarz. It is concerned with the lengths of arcs joining two given points having a curvature bounded from the above by a fixed constant. The statement of Schwarz's theorem is as follows:

THEOREM: *Let C be an arc joining two given points A and B, with curvature $k(s) \leqq 1/R$, such that $R \geqq \frac{1}{2}d$, where $d = \overline{AB}$. Let S be a circle of radius R through A and B. Then the length of C is either less than, or equal to, the shorter arc AB or greater than, or equal to, the longer arc AB on S.*

Proof: We remark that the assumption $R \geqq \frac{1}{2}d$ is necessary for the circle S to exist. To prove the theorem, we can assume that the length L of C is less than $2\pi R$; otherwise, there is nothing to prove. We then compare C with an arc of the same length on S having a chord of length d'. The conditions of Schur's theorem are satisfied and we get $d' \leqq d$, d being the distance between A and B. Hence, L is either greater than, or equal to, the longer arc of S with the chord AB, or less than, or equal to, the shorter arc of S with the chord AB.

In particular, we can consider arcs joining A and B with curvature of $1/R$, $R \geqq d/2$. The lengths of such arcs have no upper bound, as shown by the example of a helix. They have d as a lower bound, but can be as close to d as possible. Therefore, we have an example of a minimum problem which has no solution.

Finally, we remark that Schur's theorem can be generalized to sectionally smooth curves. We give here a statement of this generalization without proof.

REMARK: *Let C and C^* be two sectionally smooth curves of the same length, such that C forms a simple convex plane curve with its chord. Referred to the arc length s from one endpoint as parameter, let $k(s)$ be the curvature of C at a regular point and $a(s)$ the angle between the oriented tangents at a vertex; denote corresponding quantities for C^* by the same notations with asterisks. Let d and d^* be the distances between the endpoints of C and C^*, respectively. Then, if*

$$k^*(s) \leqq k(s) \quad \text{and} \quad a^*(s) \leqq a(s),$$

we have $d^* \geqq d$. The equality sign holds if and only if

$$k^*(s) = k(s) \quad \text{and} \quad a^*(s) = a(s).$$

The last set of conditions does not necessarily imply that C and C^* are congruent. In fact, there are simple rectilinear polygons in space which have equal sides and equal angles, but are not congruent.

6. THE GAUSS-BONNET FORMULA

We consider the intrinsic Riemannian geometry on a surface M. To simplify calculations and without loss of generality, we suppose the metric to be given in the isothermal parameters u and v:

$$(26) \qquad ds^2 = e^{2\lambda(u,\,v)}(du^2 + dv^2).$$

The element of area is then

$$(27) \qquad dA = e^{2\lambda}\, du\, dv$$

and the area of a domain D is given by the integral

$$(28) \qquad A = \iint\limits_{D} e^{2\lambda}\, du\, dv.$$

Also, the Gaussian curvature of the surface is

$$(29) \qquad K = -e^{-2\lambda}(\lambda_{uu} + \lambda_{vv}).$$

It is well-known that the Riemannian metric defines the parallelism of Levi-Civita. To express it analytically, we write

$$(30) \qquad u^1 = u \quad \text{and} \quad u^2 = v$$

and

(31) $$ds^2 = \Sigma\, g_{ij}\, du^i\, du^j.$$

In this last formula and throughout this paragraph, our small Latin indices will range from 1 to 2 and a summation sign will mean summation over all repeated indices. From g_{ij} we introduce the g^{ij}, according to the equation

(32) $$\Sigma\, g_{ij}g^{jk} = \delta_i^k$$

and the Christoffel symbols

(33) $$\begin{cases} \Gamma_{ijk} = \dfrac{1}{2}\left(\dfrac{\partial g_{ij}}{\partial u^k} + \dfrac{\partial g_{jk}}{\partial u^i} - \dfrac{\partial g_{ik}}{\partial u^j}\right) \\[2mm] \Gamma_{ik}^j = \Sigma\, g^{jh}\Gamma_{ihk} \end{cases}$$

To a vector with the components ξ^i, the Levi-Civita parallelism defines the "covariant differential"

(34) $$D\xi^i = d\xi^i + \Sigma\, \Gamma_{jk}^i\, du^k\, \xi^j.$$

All these equations are well-known in classical Riemannian geometry following the introduction of tensor analysis. The following is a new concept. Suppose the surface M is oriented. Consider the space B of all *unit* tangent vectors of M. This space B is a three-dimensional space, because the set of all unit tangent vectors with the same origin is one-dimensional. (It is called a *fiber space*, meaning that all the unit tangent vectors with origins in a neighborhood form a space which is topologically a product space.) To a unit tangent vector $\xi = (\xi^1, \xi^2)$, let $\eta = (\eta^1, \eta^2)$ be the uniquely determined unit tangent vector, orthogonal to ξ, such that ξ and η form a positive orientation. We introduce the linear differential form

(35) $$\varphi = \sum_{1 \leq i,j \leq 2} g_{ij}D\xi^i\eta^j.$$

Then φ is well-defined in B and is usually called the *connection form*.

Because the vector ξ is a unit vector, we can write its components as follows:

(36) $$\xi^1 = e^{-\lambda}\cos\theta \quad \text{and} \quad \xi^2 = e^{-\lambda}\sin\theta.$$

Then

$$(37) \qquad \eta^1 = -e^{-\lambda} \sin \theta \quad \text{and} \quad \eta^2 = e^{-\lambda} \cos \theta.$$

Routine calculation gives

$$(38) \qquad \begin{aligned} \Gamma^1_{11} &= \Gamma^2_{12} = -\Gamma^1_{22} = \lambda_u, \\ \Gamma^1_{12} &= \Gamma^2_{22} = -\Gamma^2_{11} = \lambda_v, \end{aligned}$$

whence the important relation

$$(39) \qquad \varphi = d\theta - \lambda_v \, du + \lambda_u \, dv.$$

Its exterior derivative is therefore

$$(40) \qquad d\varphi = -K \, dA.$$

Equation (40) is perhaps the most important formula in two-dimensional local Riemannian geometry.

The connection form φ is a differential form in B. We get from φ a differential form in a subset of M, when there is defined on it a field of unit tangent vectors. For example, let C be a smooth curve on M with the arc length s and let $\xi(s)$ be a smooth unit vector field along C. Then $\varphi = \sigma \, ds$, and σ is called the *variation* of ξ along C. The vectors ξ are said to be parallel along C, if $\sigma = 0$. If ξ is everywhere tangent to C, σ is called the geodesic curvature of C. C is a geodesic of M, if along C the unit tangent vectors are parallel, that is, if its geodesic curvature is 0.

Consider a domain D of M, such that there is a unit vector field defined over D, with an isolated singularity at an interior point $p_0 \in D$. Let γ_ϵ be a circle of geodesic radius ϵ about p_0. Then, from Equation (39), the limit

$$(41) \qquad \frac{1}{2\pi} \lim_{\epsilon \to 0} \int_{\gamma_\epsilon} \varphi$$

is an integer, to be called the *index* of the vector field at p_0.

Examples of vector fields with isolated singularities are shown in Figure 4. These singularities are, respectively, (a) a source or maximum, (b) a sink or minimum, (c) a center, (d) a simple saddle point, (e) a monkey saddle, and (f) a dipole. The indices are, respectively, 1, 1, 1, −1, −2, and 2.

The Gauss-Bonnet formula is the following theorem.

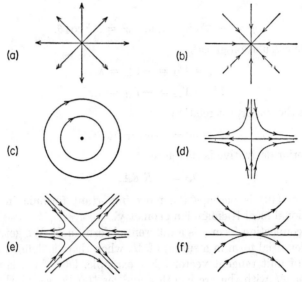

(a) (b) (c) (d) (e) (f)

FIG. 4

THEOREM: *Let D be a compact oriented domain in M bounded by a sectionally smooth curve C. Then*

$$(42) \qquad \int_C k_g \, ds + \int_D K \, dA + \sum_i (\pi - \alpha_i) = 2\pi\chi,$$

where k_g is the geodesic curvature of C, $\pi - \alpha_i$ are the exterior angles at the vertices of C, and χ is the Euler characteristic of D.

Proof: Consider first the case that D belongs to a coordinate domain (u, v) and is bounded by a simple polygon C of n sides, C_i, $1 \leqq i \leqq n$, with the angles α_i at the vertices. Suppose D is positively oriented. To the points of the arcs C_i we associate the unit tangent vectors to C_i. Thus, to each vertex is associated two vectors at an angle $\pi - \alpha_i$. By the theorem of turning tangents (see Section 1), the total variation of θ as the C_i's are traversed once is $2\pi - \sum (\pi - \alpha_i)$. It follows that

$$\int_C k_g \, ds = 2\pi - \sum_i (\pi - \alpha_i) + \int_C -\lambda_v \, du + \lambda_u \, dv.$$

By Stokes theorem, the last integral is equal to $-\iint\limits_{D} K\, dA$. Thus, the formula is proved in this special case.

In the general case, suppose D is subdivided into a union of polygons D_λ, $\lambda = 1, \cdots, f$, such that (1) each D_λ lies in one coordinate neighborhood and (2) two D_λ have either no point, or one vertex, or a whole side, in common. Moreover, let the D_λ be coherently oriented with D, so that every interior side has different senses induced by the two polygons of which it is a side. Let v and e be the numbers of interior vertices and interior sides in this subdivision of D—i.e., vertices and sides which are not on the boundary, C. The above formula can then be applied to each D_λ. Adding all these relations, we have, because the integrals of geodesic curvature along the interior sides cancel,

$$\int_C k_g\, ds + \iint\limits_D K\, dA = 2\pi f - \sum_{i,\lambda} (\pi - \alpha_{\lambda i}) - \sum_i (\pi - \alpha_i)$$

where α_i are the angles at the vertices of D, while the first sum in the right-hand side is extended over all interior vertices of the subdivision. Since each interior side is on exactly two D_λ and since the sum of interior angles about a vertex is 2π, this sum is equal to

$$-2\pi e + 2\pi v.$$

We call the integer

$$(43) \qquad \chi(D) = v - e + f$$

the Euler characteristic of D. Substituting, we get Equation (42). Equation (42) has the consequence that the integer χ is independent of the subdivision.

In particular, if C has no vertex, we have

$$(44) \qquad \int_C k_g\, ds + \iint\limits_D K\, dA = 2\pi\chi.$$

Moreover, if D is the whole surface M, we get

$$(45) \qquad \iint\limits_S K\, dA = 2\pi\chi.$$

It follows that if $K = 0$, the Euler characteristic of M is 0, and M is homeomorphic to a torus. If $K > 0$, then $\chi > 0$, and S is homeomorphic to a sphere.

The Euler characteristic plays an important role in the study of vector fields on a surface.

REMARK: *On a closed orientable surface M, the sum of the indices of a vector field with a finite number of singularities, is equal to the Euler characteristic, $\chi(M)$ of M.*

Proof: Let p_i, $1 \leq i \leq n$, be the singularities of the vector field. Let $\gamma_i(\epsilon)$ be a circle of radius ϵ about p_i, and let $\Delta_i(\epsilon)$ be the disk bounded by $\gamma_i(\epsilon)$. Integrating K of A over the domain $M - \bigcup_i \Delta_i(\epsilon)$ and using Equation (40), we get

$$\iint_{M-\bigcup_i \Delta_i(\epsilon)} K\, dA = \sum_i \int_{\gamma_i(\epsilon)} \varphi,$$

where $\gamma_i(\epsilon)$ is oriented so that it is the boundary of $\Delta_i(\epsilon)$. The theorem follows by letting $\epsilon \to 0$.

We wish to give two further applications of the Gauss-Bonnet formula. The first is a theorem of Jacobi. Let $X(s)$ be the coordinate vector of a closed space curve, with the arc length s. Let $T(s)$, $N(s)$, and $B(s)$ be the unit tangent, principal normal, and binormal vectors, respectively. In particular, the curve on the unit sphere with the coordinate vector $N(s)$ is the *principal normal indicatrix.* It has a tangent, wherever

$$(46) \qquad\qquad k^2 + w^2 \neq 0,$$

where k (when not equal to 0) and w are, respectively, the curvature and torsion of $X(s)$. Jacobi's theorem follows.

THEOREM: *If the principal normal indicatrix of a closed space curve has a tangent everywhere, it divides the unit sphere in two domains of the same area.*

Proof: To prove the theorem, we define τ by the equations

$$(47) \qquad k = \sqrt{k^2 + w^2}\cos\tau, \qquad w = \sqrt{k^2 + w^2}\sin\tau.$$

Then we have

$$d(-\cos\tau T + \sin\tau B) = (\sin\tau T + \cos\tau B)\, d\tau - \sqrt{k^2 + w^2}\, N\, ds.$$

Hence, if σ is the arc length of $N(s)$, $d\tau/d\sigma$ is the geodesic curvature of $N(s)$ on the unit sphere. Let D be one of the domains bounded by $N(s)$, and A its area. By the Gauss-Bonnet formula, we have, since $K = 1$,

$$\int_{N(s)} d\tau + \iint dA = 2\pi.$$

It follows that $A = 2\pi$, and the theorem is proved.

Our second application is Hadamard's theorem on convex surfaces.

THEOREM: *If the Gaussian curvature of a closed orientable surface in euclidean space is everywhere positive, the surface is convex (that is, it lies at one side of every tangent plane).*

We discussed a similar theorem for curves in Section 1. For surfaces, it is not necessary to suppose that it has no self-intersection.

Proof: It follows from the Gauss-Bonnet formula that the Euler characteristic $\chi(M)$ of the surface M is positive, so that $\chi(M) = 2$ and

$$\iint_M K \, dA = 4\pi.$$

Suppose M is oriented. We consider the Gauss mapping

(48) $$g: M \rightarrow \Sigma_0$$

(where Σ_0 is the unit sphere about a fixed point 0), which assigns to every point $p \in M$ the end of the unit vector through 0 parallel to the unit normal vector to M at p. The condition $K > 0$ implies that g has everywhere a nonzero functional determinant and is locally one-to-one. It follows that $g(M)$ is an open subset of Σ_0. Since M is compact, $g(M)$ is a compact subset of Σ_0, and hence is also closed. Therefore, g maps onto Σ_0.

Suppose that g is not one-to-one, that is, there exist points p and q of M, $p \neq q$, such that $g(p) = g(q)$. There is then a neighborhood U of q, such that $g(M - U) = \Sigma_0$. Since $\iint_{M-U} K \, dA$ is the area of $g(M - U)$, counted with multiplicities, we will have

$$\iint_{M-U} K \, dA \geqq 4\pi.$$

But

$$\iint_{U} K \, dA > 0,$$

so that

$$\iint_{M} K \, dA = \iint_{U} K \, dA + \iint_{M-U} K \, dA > 4\pi,$$

which is a contradiction, and Hadamard's theorem is proved.

Hadamard's theorem is true under the weaker hypothesis $K \geqq 0$, but the proof is more difficult; see the article by Chern-Lashof mentioned in Section 4.

For further reading, see:

1. S. S. Chern, "On the curvatura integra in a Riemannian manifold," *Annals of Mathematics*, 46 (1945), pp. 674–84.

2. H. Flanders, "Development of an extended exterior differential calculus," *Transactions of the American Mathematical Society*, 75 (1953), pp. 311–26.

See also Section 8 of Flanders's article in this volume.

7. UNIQUENESS THEOREMS OF COHN-VOSSEN AND MINKOWSKI

The "rigidity" theorem of Cohn-Vossen can be stated as follows.

THEOREM: *An isometry between two closed convex surfaces is established either by a motion or by a motion and a reflection.*

In other words, such an isometry is always trivial, and the theorem is obviously not true locally. The following proof is the work of G. Herglotz.

Proof: We shall first discuss some notations on surface theory in euclidean space. Let the surface S be defined by expressing its position vector X as a function of two parameters, u and v. These functions are supposed to be continuously differentiable up to the second order. Suppose that X_u and X_v are everywhere linearly independent, and let ξ be the unit normal vector, so that S is oriented. As usual, let

$$(49) \quad \begin{aligned} \mathrm{I} &= dX \cdot dX = E\, du^2 + 2F\, du\, dv + G\, dv^2 \\ \mathrm{II} &= -dX \cdot d\xi = L\, du^2 + 2M\, du\, dv + N\, dv^2 \end{aligned}$$

be the first and second fundamental forms of the surface. Let H and K denote respectively the mean and Gaussian curvatures.

It is sufficient to prove that under the isometry, the second fundamental forms are equal. Assume the local coordinates are such that corresponding points have the same local coordinates. Then E, F, and G are equal for both surfaces, and the same is true of the Christoffel symbols. Let the second surface be S^*, and denote the quantities pertaining to S^* by the same symbols with asterisks. We introduce

$$(50) \quad \lambda = \frac{L}{D}, \qquad \mu = \frac{M}{D}, \qquad \nu = \frac{N}{D},$$

where $D = \sqrt{EG - F^2}$. Then the Gaussian curvature is

$$(51) \quad K = \lambda\nu - \mu^2 = \lambda^*\nu^* - \mu^{*2},$$

and is the same for both surfaces. The mean curvatures are

$$(52) \quad H = \frac{1}{2D}\left(G\lambda - 2F\mu + E\nu\right) \quad \text{and}$$

$$H^* = \frac{1}{2D}\left(G\lambda^* - 2F\mu^* + E\nu^*\right).$$

We introduce further

$$(53) \quad J = \lambda\nu^* - 2\mu\mu^* + \nu\lambda^*.$$

The proof depends on the following identity:

$$(54) \quad DJ\xi = \frac{\partial}{\partial u}\left(\nu^* X_u - \mu^* X_v\right) - \frac{\partial}{\partial v}\left(\mu^* X_u - \lambda^* X_v\right).$$

We first notice that the Codazzi equations can be written in terms of λ^*, μ^*, and ν^* in the form

$$\lambda_v^* - \mu_u^* + \Gamma_{22}^2\lambda^* - 2\Gamma_{12}^2\mu^* + \Gamma_{11}^2\nu^* = 0,$$

(55)

$$\mu_v^* - \nu_u^* - \Gamma_{22}^1\lambda^* + 2\Gamma_{12}^1\mu^* - \Gamma_{11}^1\nu^* = 0.$$

We next write the equations of Gauss:

$$X_{uu} - \Gamma_{11}^1 X_u - \Gamma_{11}^2 X_v - D\lambda\xi = 0,$$

(56)

$$X_{uv} - \Gamma_{12}^1 X_u - \Gamma_{12}^2 X_v - D\mu\xi = 0,$$

$$X_{vv} - \Gamma_{22}^1 X_u - \Gamma_{22}^2 X_v - D\nu\xi = 0.$$

Multiplying these equations by X_v, $-X_u$, ν^*, $-2\mu^*$, and λ^*, respectively, and adding, we establish Equation (54).

We now write

(57) $$p = Xe_3, \qquad y_1 = XX_u, \qquad y_2 = XX_v,$$

where the right-hand sides are the scalar products of the vectors in question, so that $p(u, v)$ is the oriented distance from the origin to the tangent plane at $X(u, v)$. Equation (54) gives, after taking scalar product with X,

(58) $$DJp = -\nu^*E + 2\mu^*F - \lambda^*G$$
$$+ (\nu^*y_1 - \mu^*y_2)_u - (\mu^*y_1 - \lambda^*y_2)_v.$$

Let C be a closed curve on S. It divides S into two domains, D_1 and D_2, both having C as boundary. Moreover, if D_1 and D_2 are coherently oriented, C appears as a boundary in opposite senses. To each of these domains, say D_1, we apply Green's theorem, and get

(59) $$\iint_{D_1} Jp\,dA = \iint_{D_1} (-\nu^*E + 2\mu^*F - \lambda^*G)\,du\,dv$$
$$+ \int_C (+\mu^*y_1 - \lambda^*y_2)\,du + (\nu^*y_1 - \mu^*y_2)\,dv.$$

Adding this equation to a similar one for D_2, the line integrals cancel, and we have

$$\iint_S Jp\,dA = \iint_S (-\nu^*E + 2\mu^*F - \lambda^*G)\,du\,dv.$$

By Equation (52),

(60) $$\iint_S Jp\,dA = -2\iint_S H^*\,dA.$$

In particular, this formula is valid when S and S^* are identical, and we have

(61) $$\iint_S 2Kp \, dA = -2 \iint_S H \, dA.$$

Subtracting these two equations, we get

(62) $$\iint_S \begin{vmatrix} \lambda^* - \lambda & \mu^* - \mu \\ \mu^* - \mu & \nu^* - \nu \end{vmatrix} p \, dA = 2 \iint_S H^* \, dA - 2 \iint_S H \, dA.$$

To complete the proof, we need the following elementary lemma.

LEMMA: *Let*

(63) $$ax^2 + 2bxy + cy^2 \quad and \quad a'x^2 + 2b'xy + c'y^2$$

be two positive definite quadratic forms, with

(64) $$ac - b^2 = a'c' - b'^2.$$

Then

(65) $$\begin{vmatrix} a' - a & b' - b \\ b' - b & c' - c \end{vmatrix} \leqq 0,$$

and the equality sign holds only when the two forms are identical.

As proof, we observe that the statement of the lemma remains unchanged under a linear transformation of the variables. Applying such a linear transformation when necessary, we can assume $b' = b$. Then the left-hand side of Equation (65) becomes

$$(a' - a)(c' - c) = -\frac{c}{a'} (a' - a)^2 \leqq 0,$$

as to be proved. Moreover, the quantity equals 0 only when we also have $a' = a$ and $c' = c$.

We now choose the origin to be inside S, so that $p > 0$. Then the integrand in the left-hand side of Equation (62) is nonpositive, and it follows that

$$\iint_S H^* \, dA \leqq \iint_S H \, dA.$$

Since the relation between S and S^* is symmetrical, we must also have

$$\iint\limits_{S} H\, dA \le \iint\limits_{S} H^*\, dA.$$

Hence,

$$\iint\limits_{S} H\, dA = \iint\limits_{S} H^*\, dA.$$

It follows that the integral at the left-hand side of Equation (62) is 0, and hence, that

$$\lambda^* = \lambda, \qquad \mu^* = \mu, \qquad \nu^* = \nu,$$

completing the proof of Cohn-Vossen's theorem.

By Hadamard's theorem, we see that the Gauss map $g:S \to \Sigma_0$ (see Section 6) is one-to-one for a closed surface with $K > 0$. A point on S can therefore be regarded as a function of its normal vector ξ, and the same is true with any scalar function on S. Minkowski's theorem expresses the unique determination of S when $K(\xi)$ is known.

THEOREM: *Let S be a closed convex surface with Gaussian curvature $K > 0$. The function $K(\xi)$ determines S up to a translation.*

Proof: We shall give a proof of this theorem modeled after the above—that is, by an integral formula [see S. S. Chern, *American Journal of Mathematics* 79 (1957), pp. 949–50]. Let u and v be isothermal parameters on the unit sphere Σ_0, so that we have

(66) $\qquad \xi_u^2 = \xi_v^2 = A > 0 \text{ (say)}, \qquad \xi_u \xi_v = 0.$

Through the mapping g^{-1} we regard u and v also as parameters on S. Since ξ_u and ξ_v are orthogonal to ξ and are linearly independent, every vector orthogonal to ξ can be expressed as their linear combination. This fact, taken with the relation $X_u \xi_v = X_v \xi_u$, allows us to write

(67)
$$-X_u = a\xi_u + b\xi_v,$$
$$-X_v = b\xi_u + c\xi_v.$$

Forming scalar products of these equations with ξ_u and ξ_v, we have

(68) $\qquad Aa = L, \qquad Ab = M, \qquad Ac = N.$

Moreover, taking the vector product of the two relations in Equation (67), we find

$$X_u \times X_v = (ac - b^2)(\xi_u \times \xi_v).$$

But

(69) $$X_u \times X_v = D\xi, \qquad \xi_u \times \xi_v = A\xi,$$

so that, combining with Equation (68), we have

$$D = A(ac - b^2) = \frac{KD^2}{A},$$

which gives

(70) $$A = KD, \qquad ac - b^2 = \frac{1}{K}.$$

Since $A\, du\, dv$ and $D\, du\, dv$ are, respectively, the volume elements of Σ_0 and S, the first relation in Equation (70) expresses the well-known fact that K is the ratio of these volume elements.

Suppose S^* is another convex surface with the same function, $K(\xi)$. We set up a homeomorphism between S and S^*, so that they have the same normal vector at corresponding points. Then the parameters u and v can be used for both S and S^*, and corresponding points have the same parameter values. We denote by asterisks the vectors and functions for the surface S^*. Since $K = K^*$, we have from Equation (70), $ac - b^2 = a^*c^* - b^{*2}$ and $D = D^*$.

Let

(71) $$p = X \cdot \xi \quad \text{and} \quad p^* = X^* \cdot \xi$$

be the distances from the origin to the tangent planes of the two surfaces. Our basic relation is the identity

$$(X, X^*, X_u)_v - (X, X^*, X_v)_u$$

$$= A\{2(ac - b^2)p^* + (-ac^* - a^*c + 2bb^*)p\}$$

$$= A\left\{2(ac - b^2)(p^* - p) + \begin{vmatrix} a - a^* & b - b^* \\ b - b^* & c - c^* \end{vmatrix} p\right\}$$

which follows immediately from Equations (67), (69), (70), and (71). From it, we find, by Green's theorem, the integral formula

(72) $\int_{\Sigma_0} \left\{ 2(ac - b^2)(p^* - p) + \begin{vmatrix} a - a^* & b - b^* \\ b - b^* & c - c^* \end{vmatrix} p \right\} A \, du \, dv = 0.$

By translations if necessary, we can suppose the origin to be inside both surfaces, S and S^*, so that $p > 0$ and $p^* > 0$. Since

$$\begin{pmatrix} a & b \\ b & c \end{pmatrix} \quad \text{and} \quad \begin{pmatrix} a^* & b^* \\ b^* & c^* \end{pmatrix}$$

are positive definite matrices, it follows from our algebraic lemma that

$$\begin{vmatrix} a - a^* & b - b^* \\ b - b^* & c - c^* \end{vmatrix} \leq 0.$$

Hence,

(73) $\int_{\Sigma_0} (ac - b^2)(p^* - p) A \, du \, dv \geq 0.$

But the same relation is true when S and S^* are interchanged. Hence, the integral at the left-hand side of Equation (73) must be identically 0. It follows from Equation (72) that

$$\int_{\Sigma_0} \begin{vmatrix} a - a^* & b - b^* \\ b - b^* & c - c^* \end{vmatrix} p \, A \, du \, dv = 0,$$

possible only when $a = a^*, b = b^*,$ and $c = c^*$. The latter implies that

$$X_u^* = X_u \quad \text{and} \quad X_v^* = X_v,$$

which means that S and S^* differ by a translation.

For further reading, see:

1. S. S. Chern, "Integral formulas for hypersurfaces in euclidean space and their applications to uniqueness theorems," *Journal of Mathematics and Mechanics*, 8 (1959), pp. 947–55.

2. T. Otsuki, "Integral formulas for hypersurfaces in a Riemannian manifold and their applications," *Tôhoku Mathematical Journal*, 17 (1965), pp. 335–48.

3. K. Voss, "Differentialgeometrie geschlossener Flächen im euklidischen Raum," *Jahresberichte deutscher Math. Verein.*, 63 (1960–1961), pp. 117–36.

8. BERNSTEIN'S THEOREM
ON MINIMAL SURFACES

A minimal surface is a surface which locally solves the Plateau problem—that is, it is the surface of smallest area bounded by a given closed space curve. Analytically, it is defined by the condition that the mean curvature is identically 0. We suppose the surface to be given by

$$(74) \qquad z = f(x, y),$$

where the function $f(x, y)$ is twice continuously differentiable. Then a minimal surface is characterized by the partial differential equation,

$$(75) \qquad (1 + q^2)r - 2pqs + (1 + p^2)t = 0,$$

where

$$(76) \quad p = \frac{\partial f}{\partial x}, \qquad q = \frac{\partial f}{\partial y}, \qquad r = \frac{\partial^2 f}{\partial x^2}, \qquad s = \frac{\partial^2 f}{\partial x \partial y}, \qquad t = \frac{\partial^2 f}{\partial y^2}.$$

Equation (75), called the minimal surface equation, is a nonlinear "elliptic" differential equation.

Bernstein's theorem is the following "uniqueness theorem."

THEOREM: *If a minimal surface is defined by Equation (74) for all values of x and y, it is a plane. In other words, the only solution of Equation (75) valid in the whole (x, y)-plane is a linear function.*

Proof: We shall derive this theorem as a corollary of the following theorem of Jörgens [*Math Annalen* 127 (1954), pp. 130–34].

THEOREM: *Suppose the function $z = f(x, y)$ is a solution of the equation*

$$(77) \qquad rt - s^2 = 1, \quad r > 0,$$

for all values of x and y. Then $f(x, y)$ is a quadratic polynomial in x and y.

For fixed (x_0, y_0) and (x_1, y_1), consider the function

$$h(t) = f(x_0 + t(x_1 - x_0), y_0 + t(y_1 - y_0)).$$

We have

$$h'(t) = (x_1 - x_0)p + (y_1 - y_0)q,$$
$$h''(t) = (x_1 - x_0)^2 r + 2(x_1 - x_0)(y_1 - y_0)s + (y_1 - y_0)^2 s \geqq 0,$$

where the arguments in the functions p, q, r, s, t are $x_0 + t(x_1 - x_0)$ and $y_0 + t(y_1 - y_0)$. From the last inequality, it follows that

$$h'(1) \geqq h'(0)$$

or

$$(78) \qquad (x_1 - x_0)(p_1 - p_0) + (y_1 - y_0)(q_1 - q_0) \geqq 0.$$

where

$$(79) \qquad p_i = p(x_i, y_i) \quad \text{and} \quad q_i = q(x_i, y_i), \quad i = 0, 1.$$

Consider the transformation of Lewy:

$$(80) \quad \xi = \xi(x, y) = x + p(x, y), \qquad \eta = \eta(x, y) = y + q(x, y).$$

Setting

$$(81) \qquad \xi_i = \xi(x_i, y_i), \qquad \eta_i = \eta(x_i, y_i), \qquad i = 0, 1,$$

we have, by Equation (78),

$$(82) \qquad (\xi_1 - \xi_0)^2 + (\eta_1 - \eta_0)^2 \geqq (x_1 - x_0)^2 + (y_1 - y_0)^2.$$

Hence, the mapping

$$(83) \qquad\qquad\qquad (x, y) \rightarrow (\xi, \eta)$$

is distance-increasing.

Moreover, we have

$$(84) \qquad\qquad \xi_x = 1 + r, \qquad \xi_y = s$$
$$\eta_x = s, \qquad\qquad \eta_y = 1 + t,$$

so that

$$(85) \qquad\qquad \frac{\partial(\xi, \eta)}{\partial(x, y)} = 2 + r + t \geqq 2,$$

and the mapping in Equation (83) is locally one-to-one. It follows easily that Equation (83) is a diffeomorphism of the (x, y)-plane onto the (ξ, η)-plane.

We can therefore regard the function $f(x, y)$, which is a solution of Equation (77), as a function in ξ and η. Let

(86) $$F(\xi, \eta) = x - iy - (p - iq),$$

(87) $$\zeta = \xi + i\eta.$$

It can be verified by a computation that $F(\xi, \eta)$ satisfies the Cauchy-Riemann equations, so that $F(\zeta) = F(\xi, \eta)$ is a holomorphic function in ζ. Moreover, we have

(88) $$F'(\zeta) = \frac{t - r + 2is}{2 + r + t}.$$

From the last relation, we get

$$1 - |F'(\zeta)|^2 = \frac{4}{2 + r + t} > 0.$$

Thus $F'(\zeta)$ is bounded in the whole ζ-plane. By Liouville's theorem, we have

$$F'(\zeta) = \text{const.}$$

On the other hand, by Equation (88) we have

(89) $$r = \frac{|1 - F'|^2}{1 - |F'|^2}, \qquad s = \frac{i(\overline{F'} - F')}{1 - |F'|^2}, \qquad t = \frac{|1 + F'|^2}{1 - |F'|^2}.$$

It follows that r, s, and t are all constants, and Jörgens's theorem is proved.

Bernstein's theorem is an easy consequence of Jörgens's theorem. In fact, let

(90) $$W = (1 + p^2 + q^2)^{1/2}.$$

Then the minimal surface equation is equivalent to each of the following equations:

(91) $$\frac{\partial}{\partial x} \frac{-pq}{W} + \frac{\partial}{\partial y} \frac{1 + p^2}{W} = 0,$$

$$\frac{\partial}{\partial x} \frac{1 + q^2}{W} + \frac{\partial}{\partial y} \frac{-pq}{W} = 0.$$

It follows that there exists a C^2-function, $\varphi(x, y)$, such that

(92) $$\varphi_{xx} = \frac{1}{W}(1 + p^2), \qquad \varphi_{xy} = \frac{1}{W}pq, \qquad \varphi_{yy} = \frac{1}{W}(1 + q^2).$$

These partial derivatives satisfy the equation

$$\varphi_{xx}\varphi_{yy} - \varphi_{xy}^2 = 1, \quad \varphi_{xx} > 0.$$

By Jörgens's theorem, φ_{xx}, φ_{xy}, and φ_{yy} are constants. Hence, p and q are constants, and $f(x, y)$ is a linear function [This proof of Bernstein's theorem is that of J. C. C. Nitsche, *Annals of Mathematics*, 66 (1957), pp. 543–44.]

Minimal surfaces have an extensive literature. See the following expository article:

1. J. C. C. Nitsche, "On new results in the theory of minimal surfaces," *Bulletin of the American Mathematical Society*, 71 (1965), pp. 195–270.

DIFFERENTIAL FORMS

Harley Flanders

1. INTRODUCTION

It is the purpose of this exposition to discuss the calculus of differential forms and several aspects of differential geometry in which differential forms are a natural tool. Perhaps it will provide a sufficient introduction to differential geometry so that the student will pursue the subject in one of the many modern texts now available, some of which we list here.

1. Auslander, L., and R. E. MacKenzie, *Introduction to Differentiable Manifolds.* New York: McGraw-Hill Book Company, 1963.

2. Bishop, R. L., and R. J. Crittenden, *Geometry of Manifolds.* New York: Academic Press Inc., 1964.

3. Chevalley, C., *Theory of Lie Groups.* Princeton, N.J.: Princeton University Press, 1946.

4. Flanders, H., *Differential Forms with Applications to the Physical Sciences.* New York: Academic Press Inc., 1963.

5. Goldberg, S. I., *Curvature and Homology*. New York: Academic Press Inc., 1962.

6. Guggenheimer, H. W., *Differential Geometry*. New York: McGraw-Hill Book Company, 1963.

7. Helgason, S., *Differential Geometry and Symmetric Spaces*. New York: Academic Press Inc., 1962.

8. Kobayashi, S., and K. Nomizu, *Foundations of Differential Geometry* I. New York: John Wiley & Sons, Inc., 1963.

9. Lang, S., *Introduction to Differentiable Manifolds*. New York: John Wiley & Sons, Inc., 1962.

10. Lichnerowicz, A., "Théorie globale des connexions et des groupes d'holonomie," *Consiglio Nazionale delle Ricerche*. Rome: Ed. Cremonese, 1955.

11. Lichnerowicz, A., *Géométrie des groupes de transformations*. Paris: Dunod, 1958.

12. Rham, G. de, "Variétés différentiables, formes, courants, formes harmoniques," *Act. Sci. et Ind.* 1222. Paris: Hermann, 1955.

13. Sternberg, S., *Lectures on Differential Geometry*. Englewood Cliffs, N.J.: Prentice-Hall, Inc., 1964.

14. Weil, A., "Introduction à l'étude des variétés kählériennes," *Act. Sci. et Ind.* 1267. Paris: Hermann, 1958.

15. Whitney, H., *Geometric Integration Theory*. Princeton, N.J.: Princeton University Press, 1957.

16. Willmore, T. J., *Introduction to Differential Geometry*. Oxford, England: Oxford University Press, 1959.

17. Yano, K., and S. Bochner, *Curvature and Betti Numbers*. Princeton, N. J.: Princeton University Press, 1953.

Necessarily there must be a certain amount of overlap between this paper and Reference 4; most of what we shall discuss is covered in more detail there. Since we do not want simply to rehash the material of that book, we shall emphasize a number of explicit examples.

Much of modern differential geometry has its origin in classical mathematics. It is always important for the student to know the sources of the material he studies today. For this reason, we list a few very good sources of differential geometry.

18. Blaschke, W., *Vorlesungen über Differentialgeometrie* I. Berlin: Springer, 1924.

19. Blaschke, W., *Einführung in die Differentialgeometrie*. Berlin: Springer, 1950.

20. Darboux, G., *Théorie des surfaces* I–IV. Paris: Gauthier-Villars, 1887–1896.

Without any question, the most important geometer of modern times was Élie Cartan. He developed differential forms into a powerful tool in differential geometry, and his books and papers will inspire mathematical research for years to come. His pertinent work follows:

21. Cartan, É., *Leçons sur la géométrie des espaces de Riemann*. Paris: Gauthier-Villars, 1951.

22. Cartan, É., *Les systèmes différentiels extérieurs et leurs applications géométriques*. Paris: Hermann, 1945.

23. Cartan, É., *Leçons sur les invariants intégraux*. Paris: Hermann, 1922, 1958.

24. Cartan, É., *Théorie des groupes finis et continus et la géométrie différentielle*. Paris: Gauthier-Villars, 1951.

25. Cartan, É., *Oeuvres complètes*, 6 vols. Paris: Gauthier-Villars, 1952–1955.

2. DIFFERENTIAL FORMS IN $\mathbf{E}^n$

In this section, we shall not worry too much about technical details, but try to develop a feel for the object, a differential form, in $\mathbf{E}^n$. Let $(x^1, \cdots, x^n)$ denote Cartesian coordinates of a point in $\mathbf{E}^n$. For the moment, a *differential form* ω *of degree* p is the expression

$$\omega = \sum A_{i_1 \cdots i_p}(x^1, \cdots, x^n) \, dx^{i_1} \wedge dx^{i_2} \wedge \cdots \wedge dx^{i_p},$$

which occurs under the integral sign

$$\int_{\mathbf{c}^p} \omega,$$

where $\mathbf{c}^p$ is a piece of p-dimensional surface in $\mathbf{E}^n$. The familiar line integral

$$\int_{\mathbf{c}^1} (A_1 \, dx^1 + A_2 \, dx^2 + A_3 \, dx^3)$$

and surface integral

$$\int_{\mathbf{c}^2} (B_{23} \, dx^2 \wedge dx^3 + B_{31} \, dx^3 \wedge dx^1 + B_{12} \, dx^1 \wedge dx^2),$$

which we first learn in second year calculus and use so frequently in physics, are examples in $\mathbf{E}^3$.

Why is the wedge, $\wedge$, placed between the differentials? First, our integrals are always <u>oriented</u> integrals. We remember that in physics, on our piece of surface we always have an outward normal; reverse the normal and the integral changes sign. Likewise, a curve has a sense of direction the reversal of which changes the sign of the line integral.

Second, we want the rule for change of variable to be valid for these oriented integrals, which says, for example,

$$(2.1) \qquad dx^1 \wedge dx^2 = \frac{\partial(x^1, \, x^2)}{\partial(u^1, \, u^2)} \, du^1 \wedge du^2.$$

Since the Jacobian changes sign when its rows are equal, we are forced to write,

$$(2.2) \qquad dx^2 \wedge dx^1 = -dx^1 \wedge dx^2.$$

Thus, the wedge, or *exterior product*, of linear differential forms is a skew-symmetric (or alternating) product. We are also forced to have

$$(2.3) \qquad dx \wedge dx = 0,$$

because the Jacobian vanishes when its rows are equal. (For more motivation, see Flanders, pp. 1 and 2, and Sternberg, pp. 97–99.)

In $\mathbf{E}^4$ we have: 0-forms, or functions,

$$f = f(x^1, \cdots, x^4) \quad \text{(also called } scalars\text{)};$$

1-forms, or linear forms,

$$A_1 \, dx^1 + A_2 \, dx^2 + A_3 \, dx^3 + A_4 \, dx^4 \quad \text{(also called } Pfaffians\text{)};$$

2-forms,

$$B_{12} \, dx^1 \wedge dx^2 + B_{13} \, dx^1 \wedge dx^3 + \cdots + B_{34} \, dx^3 \wedge dx^4;$$

3-forms,

$$C_1 \, dx^2 \wedge dx^3 \wedge dx^4 + \cdots + C_4 \, dx^1 \wedge dx^2 \wedge dx^3;$$

and 4-forms,

$$D \, dx^1 \wedge dx^2 \wedge dx^3 \wedge dx^4 \quad \text{(also called } densities\text{)}.$$

Note that a 0-form involves one function, a 1-form involves 4 functions, a 2-form involves $\binom{4}{2} = 6$ functions, a 3-form involves 4 functions, and a 4-form involves 1 function.

Since we must soon discuss differentiation of differential forms, we shall simplify matters by assuming that the functions involved have derivatives of arbitrarily high order—that is, they are functions of class C^∞ (infinitely differentiable).

In $\mathbf{E}^n$, we use the relations (2.2) and (2.3) to express the general p-form as

$$(2.4) \qquad \omega = \sum_{1 \leq i_1 < i_2 < \cdots < i_p \leq n} A_{i_1 \cdots i_p} \, dx^{i_1} \wedge \cdots \wedge dx^{i_p},$$

so that there are

$$\binom{n}{p} = \frac{n!}{p!(n-p)!}$$

functions $A_{(i)}$ involved. An alternative expression is

$$(2.5) \qquad \omega = \frac{1}{p!} \sum A_{i_1 \cdots i_p} \, dx^{i_1} \wedge \cdots \wedge dx^{i_p},$$

where $A_{i_1 \cdots i_p}$ is skew-symmetric—that is, it vanishes whenever two indices are equal and changes sign whenever two indices are transposed.

The *exterior product* of any two differential forms is defined:

$$(2.6) \quad (\textstyle\sum A_{i_1 \cdots i_p} \, dx^{i_1} \wedge \cdots \wedge dx^{i_p}) \wedge (\sum B_{j_1 \cdots j_q} \, dx^{j_1} \wedge \cdots \wedge dx^{j_q})$$

$$= \sum (A_{i_1 \cdots} B_{j_1 \cdots})(dx^{i_1} \wedge \cdots \wedge dx^{i_p} \wedge dx^{j_1} \wedge \cdots \wedge dx^{j_q}).$$

The rules in Equations (2.2) and (2.3) are used to wipe out any terms which have a repetition of dx's and to arrange the dx's in the remaining terms in a suitable order and then to collect coefficients. (Note that $\mathbf{E}^n$ does not have forms of degree greater than n, so that if α is a p-form and β is a q-form in $\mathbf{E}^n$, and $p + q > n$, then, automatically, $\alpha \wedge \beta = 0$.)

In $\mathbf{E}^3$ we have

$$(2.7) \quad (A_1 \, dx_1 + A_2 \, dx^2 + A_3 \, dx^3) \wedge (B_1 \, dx^1 + B_2 \, dx^2 + B_3 \, dx^3)$$
$$= (A_2 B_3 - A_3 B_2) \, dx^2 \wedge dx^3 + (A_3 B_1 - A_1 B_3) \, dx^3 \wedge dx^1$$
$$+ (A_1 B_2 - A_2 B_1) \, dx^1 \wedge dx^2,$$

and

$$(2.8) \quad (A_1 \, dx^1 + A_2 \, dx^2 + A_3 \, dx^3)$$
$$\wedge (C_1 \, dx^2 \wedge dx^3 + C_2 \, dx^3 \wedge dx^1 + C_3 \, dx^1 \wedge dx^2)$$
$$= (A_1 C_1 + A_2 C_2 + A_3 C_3) \, dx^1 \wedge dx^2 \wedge dx^3.$$

Let us look at one example in $\mathbf{E}^4$. If

$$\omega = dx^1 \wedge dx^2 + dx^3 \wedge dx^4,$$

then

$$\omega \wedge \omega = 2 \, dx^1 \wedge dx^2 \wedge dx^3 \wedge dx^4.$$

In general, exterior multiplication is associative and is distributive with respect to addition. It is anticommutative in precisely this sense: If α is a p-form and β is a q-form, then

$$(2.9) \qquad\qquad \beta \wedge \alpha = (-1)^{pq} \alpha \wedge \beta.$$

In $\mathbf{E}^3$, Stokes' theorem says that if $\mathbf{c}^2$ is a piece of oriented surface the boundary curve of which, $\mathbf{c}^1 = \partial \mathbf{c}^2$, is directed accordingly, then

$$(2.10) \quad \int_{\partial \mathbf{c}^2} (A_1 \, dx^1 + A_2 \, dx^2 + A_3 \, dx^3)$$

$$= \int_{\mathbf{c}^2} \left[\left(\frac{\partial A_3}{\partial x^2} - \frac{\partial A_2}{\partial x^3} \right) dx^2 \wedge dx^3 + \left(\frac{\partial A_1}{\partial x^3} - \frac{\partial A_3}{\partial x^1} \right) dx^3 \wedge dx^1 \right.$$

$$\left. + \left(\frac{\partial A_2}{\partial x^1} - \frac{\partial A_1}{\partial x^2} \right) dx^1 \wedge dx^2 \right].$$

Thus, with each 1-form,

$$\omega = A_1\, dx^1 + A_2\, dx^2 + A_3\, dx^3,$$

there is associated a 2-form,

$$(2.11) \qquad d\omega = \left(\frac{\partial A_3}{\partial x^2} - \frac{\partial A_2}{\partial x^3}\right) dx^2 \wedge dx^3 + \cdots,$$

so that the formula

$$(2.12) \qquad \int_{\partial \mathbf{c}^2} \omega = \int_{\mathbf{c}^2} d\omega$$

is valid for each piece of oriented surface, $\mathbf{c}^2$.

One of the beauties of the calculus of exterior forms is that it is possible to define for each p-form ω a $(p + 1)$-form $d\omega$ called the *exterior derivative* of ω in such a way that Stokes' theorem goes over to arbitrary dimension: If $\mathbf{c}^{p+1}$ is an oriented $(p + 1)$-dimensional domain of integration in $\mathbf{E}^n$, and if $\mathbf{c}^p - \partial \mathbf{c}^{p+1}$ is its p-dimensional boundary (also a domain of integration) which is oriented in a way coherent to the orientation of $\mathbf{c}^{p+1}$, then

$$(2.13) \qquad \int_{\partial \mathbf{c}^{p+1}} \omega = \int_{\mathbf{c}^{p+1}} d\omega.$$

This exterior derivative d has a very simple definition. It is additive, so that we need only write it down for an individual summand in Equation (2.4):

$$(2.14) \quad d(A\, dx^{i_1} \wedge \cdots \wedge dx^{i_p})$$

$$= \left(\sum \frac{\partial A}{\partial x^i} dx^i\right) \wedge (dx^{i_1} \wedge \cdots \wedge dx^{i_p}).$$

The exterior derivative has the following characteristic properties:

$$(2.15) \quad \begin{cases} dA = \sum \dfrac{\partial A}{\partial x^i} dx^i \quad \text{for a scalar } A; \\[2mm] d(\alpha \wedge \beta) = (d\alpha) \wedge \beta + (-1)^p \alpha \wedge (d\beta) \\[1mm] \qquad\qquad\qquad\qquad\qquad \text{if } \alpha \text{ is a } p\text{-form;} \\[2mm] d(d\alpha) = 0 \quad \text{for any } \alpha. \end{cases}$$

Let us note in passing another familiar case of the general Stokes' theorem (2.13). We let $\mathbf{c}^3$ be a three-dimensional domain of integration in $\mathbf{E}^3$. Then,

$$(2.16) \quad \int_{\partial c^3} (A_1\, dx^2 \wedge dx^3 + A_2\, dx^3 \wedge dx^1 + A_3\, dx^1 \wedge dx^2)$$

$$= \int_{c^3} \left(\frac{\partial A_1}{\partial x^1} + \frac{\partial A_2}{\partial x^2} + \frac{\partial A_3}{\partial x^3} \right) dx^1 \wedge dx^2 \wedge dx^3,$$

which is the theorem of Ostrogradsky (or Gauss).

For practice, let us compute the exterior derivative of a particular 2-form in $\mathbf{E}^4$:

$$d(A\, dx^1 \wedge dx^2 + B\, dx^1 \wedge dx^4)$$

$$= (dA) \wedge dx^1 \wedge dx^2 + (dB) \wedge dx^1 \wedge dx^4$$

$$= \left(\frac{\partial A}{\partial x^1} dx^1 + \frac{\partial A}{\partial x^2} dx^2 + \frac{\partial A}{\partial x^3} dx^3 + \frac{\partial A}{\partial x^4} dx^4 \right) \wedge dx^1 \wedge dx^2$$

$$+ \left(\frac{\partial B}{\partial x^1} dx^1 + \frac{\partial B}{\partial x^2} dx^2 + \frac{\partial B}{\partial x^3} dx^3 + \frac{\partial B}{\partial x^4} dx^4 \right) \wedge dx^1 \wedge dx^4$$

$$= \frac{\partial A}{\partial x^3} dx^1 \wedge dx^2 \wedge dx^3 + \left(\frac{\partial A}{\partial x^4} - \frac{\partial B}{\partial x^2} \right) dx^1 \wedge dx^2 \wedge dx^4$$

$$- \frac{\partial B}{\partial x^3} dx^1 \wedge dx^3 \wedge dx^4.$$

The last formula in Equations (2.15), $d(d\alpha) = 0$, is known as the Poincaré lemma. It is very important, first because it allows us to think of d as a coboundary operation with topological implications, and second, because it is the source of what are called integrability conditions in differential geometry. This relation, $d(d\alpha) = 0$, is nothing more than the equality of mixed second partials:

$$\frac{\partial^2 A}{\partial x \partial y} = \frac{\partial^2 A}{\partial y \partial x}.$$

(For details on the exterior derivative, see Flanders, pp. 20–22, and Sternberg, pp. 99–101; for Stokes' theorem, see Flanders, pp. 64–66.)

From Equations (2.7), (2.8), (2.10), and (2.16), we see that the calculus of exterior differential forms includes the standard operations of vector analysis: vector product, scalar product, curl (rotation), and divergence.

We come now to the most striking property of differential forms —their behavior under mappings. We take two euclidean spaces—

$\mathbf{E}^m$ with coordinates $x^1, \cdots, x^m$, and $\mathbf{E}^n$ with coordinates $y^1, \cdots, y^n$. We suppose given an infinitely differentiable mapping,

$$\phi : \mathbf{E}^m \to \mathbf{E}^n,$$

which we may express in coordinates by

$$(2.17) \qquad y^i = y^i(x^1, \cdots, x^m) \quad (i = 1, \cdots, n).$$

If

$$\omega = \Sigma \, A_{i_1 \cdots i_p}(y^1, \cdots, y^n) \, dy^{i_1} \wedge \cdots \wedge dy^{i_p}$$

is any p-form on $\mathbf{E}^n$, we may associate with it an *induced* p-form $\phi^*\omega$ on $\mathbf{E}^m$ by the following substitution rules. In the expression for ω, replace each coefficient $A(y^1, \cdots, y^n)$ by the composite function,

$$A(y^1(x^1, \cdots, x^m), \cdots, y^n(x^1, \cdots, x^m))$$

and replace each dy^i by the linear differential,

$$\frac{\partial y^i}{\partial x^1} \, dx^1 + \cdots + \frac{\partial y^i}{\partial x^m} \, dx^m;$$

then use exterior multiplication and consolidate terms.

Thus, from the mapping $\phi : \mathbf{E}^m \to \mathbf{E}^n$, we have constructed a mapping ϕ^* which takes each p-form ω on $\mathbf{E}^n$ to a p-form $\phi^*\omega$ on $\mathbf{E}^m$. The operations on differential forms are preserved in precisely this sense:

$$(2.18) \qquad \begin{cases} \phi^*(\omega_1 + \omega_2) = \phi^*\omega_1 + \phi^*\omega_2, \\ \phi^*(\omega \wedge \eta) = (\phi^*\omega) \wedge (\phi^*\eta), \\ \phi^*(d\omega) = d(\phi^*\omega). \end{cases}$$

Proving these relations is an excellent test of our grasp of this subject so far; of course, we may read proofs in any of the books which discusses exterior forms.

If, in addition to $\phi : \mathbf{E}^m \to \mathbf{E}^n$, we also have $\psi : \mathbf{E}^n \to \mathbf{E}^r$, then $\psi \circ \phi : \mathbf{E}^m \to \mathbf{E}^r$. If ω is a p-form on $\mathbf{E}^r$, then we have

$$(2.19) \qquad (\psi \circ \phi)^*\omega = \phi^*(\psi^*\omega),$$

which we may write simply as

$$(2.20) \qquad (\psi \circ \phi)^* = \phi^* \circ \psi^*.$$

On checking this, we shall recognize that it is essentially the chain rule.

3. THE POINCARÉ LEMMA

The Poincaré lemma, $d(d\alpha) = 0$, may be stated as follows.

LEMMA: *If ω is a p-form on $\mathbf{E}^n$ for which there exists a $(p - 1)$-form α such that $d\alpha = \omega$, then $d\omega = 0$.*

It is remarkable and not at all obvious that the converse is true.

LEMMA (Converse):

If ω is a p-form on $\mathbf{E}^n$ such that $d\omega = 0$, then there is a $(p - 1)$-form α such that $\omega = d\alpha$. (Exception: $p = 0$. Then $\omega = f$ is a scalar and the vanishing of df simply means f is constant.)

The standard proof of a cylinder construction motivated by topology is given, for example, in Flanders, pp. 27–29. A more direct proof is given in W. V. D. Hodge, *Harmonic Integrals*, pp. 94–97, and in an elegant form in Sternberg, pp. 121–23. We shall here work out one particular case by direct calculation, because it will illustrate what the result means. (We have now come far enough so that we may save writing by dropping the wedge, $\wedge$. Henceforth, we understand $dx\,dy$ means $dx \wedge dy$, etc.)

Proof: We work in $\mathbf{E}^4$ and suppose ω is a 2-form such that $d\omega = 0$.

Thus,

$$(3.1) \quad \omega = P_{12}\,dx^1\,dx^2 + P_{13}\,dx^1\,dx^3 + P_{14}\,dx^1\,dx^4$$
$$+ P_{23}\,dx^2\,dx^3 + P_{24}\,dx^2\,dx^4 + P_{34}\,dx^3\,dx^4.$$

The condition $d\omega = 0$ gives us a system of 4 equations:

$$(3.2) \quad \begin{cases} \dfrac{\partial P_{12}}{\partial x^3} - \dfrac{\partial P_{13}}{\partial x^2} + \dfrac{\partial P_{23}}{\partial x^1} = 0, \\[2mm] \dfrac{\partial P_{12}}{\partial x^4} - \dfrac{\partial P_{14}}{\partial x^2} + \dfrac{\partial P_{24}}{\partial x^1} = 0, \\[2mm] \dfrac{\partial P_{13}}{\partial x^4} - \dfrac{\partial P_{14}}{\partial x^3} + \dfrac{\partial P_{34}}{\partial x^1} = 0, \\[2mm] \dfrac{\partial P_{23}}{\partial x^4} - \dfrac{\partial P_{24}}{\partial x^3} + \dfrac{\partial P_{34}}{\partial x^2} = 0. \end{cases}$$

We seek a 1-form,

$$(3.3) \qquad \alpha = A_1\, dx^1 + A_2\, dx^2 + A_3\, dx^3 + A_4\, dx^4,$$

such that $d\alpha = \omega$. It gives us a system of 6 equations for the 4 unknown functions, $A_1, \cdots, A_4$:

$$(3.4) \quad \begin{cases} \dfrac{\partial A_2}{\partial x^1} - \dfrac{\partial A_1}{\partial x^2} = P_{12}, & \dfrac{\partial A_3}{\partial x^2} - \dfrac{\partial A_2}{\partial x^3} = P_{23}, \\[2ex] \dfrac{\partial A_3}{\partial x^1} - \dfrac{\partial A_1}{\partial x^3} = P_{13}, & \dfrac{\partial A_4}{\partial x^2} - \dfrac{\partial A_2}{\partial x^4} = P_{24}, \\[2ex] \dfrac{\partial A_4}{\partial x^1} - \dfrac{\partial A_1}{\partial x^4} = P_{14}, & \dfrac{\partial A_4}{\partial x^3} - \dfrac{\partial A_3}{\partial x^4} = P_{34}. \end{cases}$$

We note first that if α is a solution, then so is $\alpha + df$ where f is any function [since $d(\alpha + df) = d\alpha + d(df) = d\alpha$]. This situation suggests we try for a solution in which $A_1 = 0$. The first column of (3.4) becomes

$$(3.5) \qquad \frac{\partial A_2}{\partial x^1} = P_{12}, \qquad \frac{\partial A_3}{\partial x^1} = P_{13}, \qquad \frac{\partial A_4}{\partial x^1} = P_{14}.$$

By taking

$$A_2(x^1, x^2, x^3, x^4) = \int_0^{x^1} P_{12}(t, x^2, x^3, x^4)\, dt, \quad \text{etc.,}$$

we determine definite solutions of (3.5). We obtain the most general solution of (3.5) by taking arbitrary functions

$$B_2 = B_2(x^2, x^3, x^4), \qquad B_3 = B_3(x^2, x^3, x^4), \qquad B_4 = B_4(x^2, x^3, x^4),$$

and adding these to A_2, A_3, and A_4, respectively. We try to choose $B_2, \cdots$ so that the second column of (3.4) is also satisfied. It reduces to

$$(3.6) \quad \begin{cases} \dfrac{\partial B_3}{\partial x^2} - \dfrac{\partial B_2}{\partial x^3} = P_{23} - \dfrac{\partial A_3}{\partial x^2} + \dfrac{\partial A_2}{\partial x^3}, \\[2ex] \dfrac{\partial B_4}{\partial x^2} - \dfrac{\partial B_2}{\partial x^4} = P_{24} - \dfrac{\partial A_4}{\partial x^2} + \dfrac{\partial A_2}{\partial x^4}, \\[2ex] \dfrac{\partial B_4}{\partial x^3} - \dfrac{\partial B_3}{\partial x^4} = P_{34} - \dfrac{\partial A_4}{\partial x^3} + \dfrac{\partial A_3}{\partial x^4}. \end{cases}$$

Obviously, if these equations are solvable, the right-hand sides must be functions of x^2, x^3, and x^4 alone. However, for example,

$$\frac{\partial}{\partial x^1}\left(P_{23} - \frac{\partial A_3}{\partial x^2} + \frac{\partial A_2}{\partial x^3}\right) = \frac{\partial P_{23}}{\partial x^1} - \frac{\partial P_{13}}{\partial x^2} + \frac{\partial P_{12}}{\partial x^3} = 0,$$

by the first equation in (3.2). Similarly, each of the first three of the equations in (3.2) means that the right-hand side of the corresponding equation in (3.6) has vanishing partial with respect to x^1, and hence is independent of x^1.

The system in (3.6) is now similar to our original system, (3.4), but in one less variable. We proceed the same way, setting $B_2 = 0$ and determining B_3 and B_4 by the first two equations in (3.6). They are determined to addition of functions $C_3(x^3, x^4)$ and $C_4(x^3, x^4)$. We must choose these to satisfy the last equation in (3.6), which now becomes

$$(3.7) \qquad \frac{\partial C_4}{\partial x^3} - \frac{\partial C_3}{\partial x^4} = P_{34} - \frac{\partial B_4}{\partial x^3} + \frac{\partial B_3}{\partial x^4} - \frac{\partial A_4}{\partial x^3} + \frac{\partial A_3}{\partial x^4}.$$

For this equation to have a solution, the right-hand side must be independent of x^2. But

$$\frac{\partial}{\partial x^2}\left(P_{34} - \frac{\partial B_4}{\partial x^3} + \frac{\partial B_3}{\partial x^4} - \frac{\partial A_4}{\partial x^3} + \frac{\partial A_3}{\partial x^4}\right)$$

$$= \frac{\partial P_{34}}{\partial x^2} - \frac{\partial}{\partial x^3}\left(P_{24} - \frac{\partial A_4}{\partial x^2} + \frac{\partial A_2}{\partial x^4}\right) + \frac{\partial}{\partial x^4}\left(P_{23} - \frac{\partial A_3}{\partial x^2} + \frac{\partial A_2}{\partial x^3}\right)$$

$$- \frac{\partial^2 A_4}{\partial x^2 \partial x^3} + \frac{\partial^2 A_3}{\partial x^2 \partial x^4} = \frac{\partial P_{34}}{\partial x^2} - \frac{\partial P_{24}}{\partial x^3} + \frac{\partial P_{23}}{\partial x^4} = 0,$$

according to the last of the integrability conditions in (3.2). Having this, we may now take $C_3 = 0$ and C_4, an indefinite integral with respect to x^3 of the right-hand side of Equation (3.7). The proof is now complete.

This computation suggests a systematic procedure for the general case. The reader will do well to give the suggestion some thought and also to try the cases in which ω is a 3-form or a 4-form in $\mathbf{E}^4$. Note that the equation $d\omega = 0$ in the latter case says nothing ($\mathbf{E}^4$ has no 5-forms), so that we assert each 4-form ω in $\mathbf{E}^4$ is the exterior derivative of a 3-form.

4. MANIFOLDS

If we look critically at what properties of euclidean space E^n we have used to set up the calculus of differential forms, we see that the coordinate system $x^1, \cdots, x^n$ is the only one we have used. However, according to the behavior of differential forms under mappings [Equations (2.18) and (2.19)], any other coordinate system could have been used as long as the two systems were related by C^∞ functions. In particular, we automatically have a theory of differential forms on any open subset of E^n simply by restricting the euclidean coordinate functions to the open subset.

But this method is far from adequate, since we want to study geometric spaces which cannot be considered as such open subsets of E^n. For example, the ordinary 2-sphere S^2 is a two-dimensional surface in E^3 which cannot be imbedded in the plane. We need a theory of differential forms on such surfaces.

It might seem adequate to extend the definition of (nonsingular) surface in E^3 to a definition of a (nonsingular) variety of dimension p in E^n. Actually, it is sufficient, but only as a result of a deep imbedding theorem of H. Whitney. However, there are many rather simple spaces which would appear to have sufficient structure for study by analytic methods but which we do not identify in any apparent fashion with subspaces of E^n. Two examples will suffice.

Examples

1. *The space of all directed lines in* E^3. A minute's reflection will convince us that this is a four-dimensional space. It appears to be smooth—no corners, edges, etc.—and it is not obvious how to consider it as a subspace of an E^n. (The Whitney theorem just mentioned shows that this space may be considered a subspace of E^8, but does not do so explicitly. See Sternberg, p. 63, and Whitney, pp. 111ff.)†

---

† See also Whitney, "Self-intersections of a manifold," *Annals of Mathematics*, (2) 45 (1944), pp. 220–26 (esp. 236).

2. *Projective space* $\mathbf{P}^n$. This object may be considered as the set of all lines (undirected) through the origin in $\mathbf{E}^{n+1}$. Alternatively, it is obtained by identifying antipodal points on the sphere $\mathbf{S}^n$. (The projective plane, $\mathbf{P}^2$, can be imbedded in $\mathbf{E}^4$ but not in $\mathbf{E}^3$.)

Now we set down the formal definition of a (differentiable) manifold, probably the most general kind of structure on which we perform differential geometry (see Lang), and on which the differential form calculus is possible. The idea is to have a topological space each point of which has a neighborhood with coordinate functions making said neighborhood indistinguishable from a neighborhood in $\mathbf{E}^n$.

We fix our object, $\mathbf{M}$, a connected topological space, and we fix what will be its dimension, n.

DEFINITION (4.1): *A chart on* $\mathbf{M}$ *is a pair* $(\mathbf{U}, f)$ *where* $\mathbf{U}$ *is an open subset of* $\mathbf{M}$ *and* f *is a homeomorphism on* $\mathbf{U}$ *onto an open subset* $f(\mathbf{U})$ *of* $\mathbf{E}^n$.

DEFINITION (4.2): *An atlas on* $\mathbf{M}$ *is a set* $\mathfrak{A}$ *of charts such that:* (1) *the sets* $\mathbf{U}$ *coming from the charts* $(\mathbf{U}, f)$ *of* $\mathfrak{A}$ *cover* $\mathbf{M}$; (2) *if* $(\mathbf{U}, f)$ *and* $(\mathbf{V}, g)$ *are in the atlas, and if* $\mathbf{U} \cap \mathbf{V} \neq \varnothing$, *then the mapping* $g \circ f^{-1} : f(\mathbf{U} \cap \mathbf{V}) \to g(\mathbf{U} \cap \mathbf{V})$ *is of class* C^{∞}; *and* (3) $\mathfrak{A}$ *is maximal with respect to properties* (1) *and* (2).

DEFINITION (4.3): *A differentiable manifold of dimension* n *is a triple* $(\mathbf{M}, n, \mathfrak{A})$ *where* $\mathbf{M}$ *is a connected topological space, and* $\mathfrak{A}$ *is an atlas of charts* $(\mathbf{U}, f)$, $f : \mathbf{U} \to \mathbf{E}^n$.

We commonly refer to the space $\mathbf{M}$ itself as the differentiable manifold, keeping in mind the structure imposed on this space. (This is no worse than saying "the group $\mathbf{G}$" or "the topological space $\mathbf{X}$," etc.)

Let $p \in \mathbf{M}$ and let $(\mathbf{U}, f)$ be a chart such that $p \in \mathbf{U}$. Now, $f(\mathbf{U})$ is a subset of $\mathbf{E}^n$ with its Cartesian coordinates $x^1, \cdots, x^n$. Since $f(\mathbf{U})$ is homeomorphic to $\mathbf{U}$, we may think of $x^1, \cdots, x^n$ as functions on $\mathbf{U}$, in which case we get what is called a *local coordinate system* at p.

If $y^1, \cdots, y^n$ is another local coordinate system at p, we may write

$$y^i = y^i(x^1, \cdots, x^n),$$

and these functions will be infinitely differentiable where defined [according to Definition (4.2)].

A real function ϕ on $\mathbf{M}$ is *infinitely differentiable* if it is an infinitely differentiable function of the local coordinates $x^1, \cdots, x^n$ on each such $\mathbf{U}$.

Similarly, we define an infinitely differentiable mapping

$$\phi : \mathbf{M} \to \mathbf{N}$$

from one manifold to another.

Almost every one of the recent treatises we have listed in Section 1 treats seriously the basic properties of C^∞ manifolds. An interesting treatment of the real analytic case is found in Chevalley, pp. 68ff. It is important for us to define differential forms on a manifold $\mathbf{M}$.

To this end, let $\mathbf{M}$ be a manifold of dimension n with atlas $\mathfrak{A}$.

DEFINITION (4.4): *A differential form of degree p on $\mathbf{M}$ is defined by assigning to each chart $(\mathbf{U}, f)$ of $\mathfrak{A}$ a p-form*

$$\omega_{(f,\mathbf{U})} \quad \text{on } f(\mathbf{U})$$

so that whenever $(f, \mathbf{U})$ and $(g, \mathbf{V})$ are two charts with $\mathbf{U} \cap \mathbf{V} \neq \varnothing$, and

$$g \circ f^{-1} : f(\mathbf{U} \cap \mathbf{V}) \to g(\mathbf{U} \cap \mathbf{V}),$$

then

$$(g \circ f^{-1})^* \omega_{(g,\mathbf{V})} = \omega_{(f,\mathbf{U})} \quad \text{on } g(\mathbf{U} \cap \mathbf{V}).$$

In other words, we take a p-form in each local coordinate system and insist that these "pieces of a form" fit together properly. Once we have said it right, all our previous work is applicable. In particular, $\omega_1 + \omega_2$ is defined as are $\omega \wedge \eta$ and $d\omega$. We have $d(d\omega) = 0$. If

$$\phi : \mathbf{M} \to \mathbf{N},$$

then $\phi^* \omega$ is a p-form on $\mathbf{M}$ for each p-form ω on $\mathbf{N}$, and

$$(4.5) \quad \begin{cases} \phi^*(\omega_1 + \omega_2) = \phi^*\omega_1 + \phi^*\omega_2, \\ \phi^*(\omega \wedge \eta) = (\phi^*\omega) \wedge (\phi^*\eta), \\ \phi^*(d\omega) = d(\phi^*\omega). \end{cases}$$

If $\phi: \mathbf{M} \to \mathbf{N}$ and $\psi: \mathbf{N} \to \mathbf{P}$, then

(4.6) $(\psi \circ \phi)^* = \phi^* \circ \psi^*.$

All these statements are proved by what seem to the experienced to be routine applications of the definitions and the corresponding statements in Section 2. However, the beginner must work through proofs himself to be sure he understands the subject.

Differential forms on a manifold can also be defined in a more sophisticated way as cross sections of a suitable fiber bundle. (See Sternberg, pp. 77–96.)

[Note: Requirement (3) of Definition (4.2) seems puzzling at first. It is included only for the technical reason that without it we must define equivalence classes of atlases. For practical applications, an atlas is completely determined by any subset which contains enough charts $(\mathbf{U}, f)$ so that the collection of $\mathbf{U}$'s covers $\mathbf{M}$. To define a p-form, it is enough to give the $\omega_{(\mathbf{U},f)}$ for these charts of the subset, subject to the condition of fitting together, of course.

5. INTEGRATION

Let us begin with a review of line and surface integrals in $\mathbf{E}^3$. For a line integral, we are given a curve $\mathbf{c}^1$ in $\mathbf{E}^3$ and a 1-form ω defined in a domain which includes the curve. The problem is to define

$$\int_{\mathbf{c}^1} \omega.$$

The usual procedure is to break the curve into several pieces, each of which has a parameterization, to give the integral by an explicit formula for each piece, and to sum the results.

The simplest kind of parametric curve will be given by a one-one smooth mapping

(5.1) $\phi: t \to (x^1(t), \cdots, x^3(t))$

on $0 \leqq t \leqq 1$ into $\mathbf{E}^3$, such that the velocity vector,

$$\left(\frac{dx^1}{dt}, \ldots, \frac{dx^3}{dt} \right),$$

never vanishes. We then substitute into

$$\omega = A_1 \, dx^1 + A_2 \, dx^2 + A_3 \, dx^3$$

the expressions for $x^1, \cdots, x^3$ in terms of t given by Equation (5.1), and we set

$$(5.2) \qquad \int_{\mathbf{c}^1} \omega = \int_0^1 \left[A_1(x^1(t), \cdots, x^3(t)) \frac{dx^1}{dt} + \cdots \right] dt.$$

In our language,

$$(5.3) \qquad \int_{\mathbf{c}^1} \omega = \int_0^1 \phi^* \omega.$$

Let us look at it another way. We let $\mathbf{e}^1$ denote the unit interval on the t-axis, a very special parameterized curve. Its image under ϕ is $\mathbf{c}^1$ and we write

$$(5.4) \qquad \phi_* \mathbf{e}^1 = \mathbf{c}^1$$

to emphasize this relation. Then Equation (5.3) becomes

$$(5.5) \qquad \int_{\phi_* \mathbf{e}^1} \omega = \int_{\mathbf{e}^1} \phi^* \omega,$$

a suggestive formulation.

Let us examine a surface integral. We are given a 2-form,

$$\omega = P_1 \, dx^2 \, dx^3 + P_2 \, dx^3 \, dx^1 + P_3 \, dx^1 \, dx^2$$

on an open set $\mathbf{U}$ of $\mathbf{E}^3$ (again, omitting the wedge, $\wedge$). We are also given an oriented surface $\mathbf{c}^2$ in $\mathbf{U}$ and seek

$$(5.6) \qquad \int_{\mathbf{c}^2} \omega.$$

The surface $\mathbf{c}^2$ is broken up into parameterized pieces and the integrals for these pieces are summed. Thus, let us suppose $\mathbf{c}^2$ is one of these pieces—that is, $\mathbf{c}^2$ is given as the image of a smooth one-one mapping,

$$(5.7) \qquad \phi: (u^1, u^2) \to (x^1(u^1, u^2), \cdots, x^3(u^1, u^2)).$$

We assume (u^1, u^2) runs over an integration domain $\mathbf{e}^2 \subset \mathbf{E}^2$ and the mapping is regular in the sense that the two velocity vectors,

$$\left(\frac{\partial x^1}{\partial u^1}, \frac{\partial x^2}{\partial u^1}, \frac{\partial x^3}{\partial u^1}\right),$$

(5.8)

$$\left(\frac{\partial x^1}{\partial u^2}, \frac{\partial x^2}{\partial u^2}, \frac{\partial x^3}{\partial u^2}\right),$$

which correspond to motion along the parameter curves, are linearly independent at each point. The standard definition is

$$(5.9) \quad \int_{\mathbf{c}^2} \omega = \int_{\mathbf{e}^2}\left[P_1(x^1(u^1, u^2), \cdots) \frac{\partial(x^2, x^3)}{\partial(u^1, u^2)} + \cdots \right.$$

$$\left. + P_3(x^1(u^1, u^2), \cdots) \frac{\partial(x^1, x^2)}{\partial(u^1, u^2)}\right] du^1\, du^2.$$

But, for example,

$$\phi^*(dx^2\, dx^3) = \left(\frac{\partial x^2}{\partial u^1}\, du^1 + \frac{\partial x^2}{\partial u^2}\, du^2\right)\left(\frac{\partial x^3}{\partial u^1}\, du^1 + \frac{\partial x^3}{\partial u^2}\, du^2\right)$$

$$= \left(\frac{\partial x^2}{\partial u^1}\frac{\partial x^3}{\partial u^2} - \frac{\partial x^2}{\partial u^2}\frac{\partial x^3}{\partial u^1}\right) du^1\, du^2$$

$$= \frac{\partial(x^2, x^3)}{\partial(u^1, u^2)}\, du^1\, du^2, \quad \text{etc.},$$

so that Equation (5.9) is, simply,

$$(5.10) \qquad \int_{\mathbf{c}^2} \omega = \int_{\mathbf{e}^2} \phi^*\omega.$$

Finally, if we formally recognize that $\mathbf{c}^2$ is the image of $\mathbf{e}^2$ under ϕ by writing

$$(5.11) \qquad\qquad \phi_*\mathbf{e}^2 = \mathbf{c}^2,$$

we have

$$\int_{\phi_*\mathbf{e}^2} \omega = \int_{\mathbf{e}^2} \phi^*\omega.$$

These examples are given to motivate the formal theory of integrals of differential forms over chains. For a complete treatment of this subject, see Whitney (Reference 15). Treatments adequate for most of differential geometry appear in Sternberg, pp. 104–11, and in Flanders, pp. 57–66. The result of this theory is that on a manifold **M** there is a class of objects called p-chains over which one can integrate p-forms. Each $(p + 1)$-chain, $\mathbf{c}^{p+1}$, has a boundary, $\mathbf{c}^p = \partial \mathbf{c}^{p+1}$, which is a p-chain, and the general Stokes theorem [Equation (2.13)] is valid.

It is technically convenient to allow chains which cross over onto themselves and bunch up. Therefore, for the preceding examples, the mappings ϕ of Equations (5.1) and (5.7) do not have to be either one-one or regular. Because of this added freedom, it is not hard to establish the following. Let $\mathbf{M}$ and $\mathbf{N}$ be manifolds and let

$$(5.12) \qquad \phi : \mathbf{M} \to \mathbf{N}$$

be a smooth mapping. Let $\mathbf{c}^p$ be a p-chain on $\mathbf{M}$ and ω a p-form on $\mathbf{N}$. Then $\phi_* \mathbf{c}^p$ is a p-chain on $\mathbf{N}$, $\phi^* \omega$ is a p-form on $\mathbf{M}$, and

$$(5.13) \qquad \int_{\mathbf{c}^p} \phi^* \omega = \int_{\phi_* \mathbf{c}^p} \omega.$$

In the next section, we shall make some explicit computations to illustrate these matters.

6. EXAMPLES OF THE THEOREMS
OF DE RHAM

At the beginning of Section 3, we stated the converse to the Poincaré lemma, which says that on $\mathbf{E}^n$ a form is *exact* (equal to an exterior derivative) if and only if it is *closed* (its exterior derivative vanishes). This theorem is local and is not true as stated for any manifold. If $\mathbf{M}$ is a manifold, let $F^p = F^p(\mathbf{M})$ denote the real vector space of all p-forms on $\mathbf{M}$. Then

$$(6.1) \qquad d : F^p \to F^{p+1}.$$

The kernel of d, in each dimension, is the space C^p of *closed* forms. We write,

$$(6.2) \qquad C^p = \{\omega \in F^p \mid d\omega = 0\}.$$

The image of d (from the previous dimension) is the space Ex^p of *exact* forms. We write

$$(6.3) \qquad Ex^p = d(F^{p-1}).$$

Since $d(d\alpha) = 0$ for any form α, we have

$$(6.4) \qquad Ex^p \subseteq C^p.$$

The quotient space

(6.5) $$H^p = \frac{C^p}{Ex^p}$$

is the pth *de Rham cohomology group* of **M**. We know that it vanishes for $\mathbf{E}^n$. The main de Rham theorem asserts that it is naturally isomorphic (for $p > 0$) to the topological cohomology group (with real coefficients) of **M** if **M** is compact. We shall explore some examples.

Examples

1. First we take $\mathbf{M} = \mathbf{S}^1$, the *circle*. We may take the central angle θ (mod 2π) as parameter. A 1-form

$$\omega = f(\theta)\, d\theta \qquad (f(\theta + 2\pi) = f(\theta))$$

is exact if there is a periodic function g such that

$$f(\theta) = \frac{dg}{d\theta}.$$

Then

$$\int_{\mathbf{S}^1} \omega = \int_0^{2\pi} \frac{dg}{d\theta}\, d\theta = g(2\pi) - g(0) = 0,$$

suggesting that *a 1-form ω on $\mathbf{S}^1$ is exact if and only if*

(6.6) $$\int_{\mathbf{S}^1} \omega = 0.$$

We have just seen that the condition is necessary. If, on the other hand, the integral vanishes, then we may set

$$g(\theta) = \int_0^\theta f(t)\, dt$$

and, (the important point) this relation is well-defined for θ mod 2π. Clearly,

$$\frac{dg}{d\theta} = f(\theta), \qquad dg = \omega.$$

2. For our next example, we take the *cylinder*,

(6.7) $(-1, 1) \times \mathbf{S}^1 = \{(t, \theta) \mid -1 < t < 1, \quad \theta \text{ mod } 2\pi\}.$

We denote the equator by

(6.8) $$\mathbf{c}^1 = \{0\} \times \mathbf{S}^1.$$

If f is a 0-form on the cylinder and $\omega = df$, then

$$\int_{\mathbf{c}^1} \omega = \int_{\mathbf{c}^1} df = \int_{\partial \mathbf{c}^1} f = 0.$$

This relation proves the "only if" part of the following assertion.

PROPOSITION: *Let ω be a closed 1-form on the cylinder. Then ω is exact if and only if*

(6.9) $$\int_{\mathbf{c}^1} \omega = 0.$$

Proof: We could prove the sufficiency by showing that the line integral

$$\int_{P_0}^{P_1} \omega$$

is independent of the path. Instead, we shall give a different type of proof which will guide the study of the projective plane which follows. To write something definite, we shall think of $\mathbf{S}^1$ as the unit circle in the (x, y) plane so that $\theta \to (\cos \theta, \sin \theta)$ is the polar parameterization of $\mathbf{S}^1$. We consider the mapping

(6.10) $$\phi : (-1, 1) \times \mathbf{E}^1 \to (-1, 1) \times \mathbf{S}^1$$

given by

(6.11) $$\phi(t, \theta) = (t, \cos \theta, \sin \theta),$$

which gives a covering of the cylinder by the infinite strip. Now we are given a closed 1-form ω on the cylinder which satisfies the relation,

$$\int_{\mathbf{c}^1} \omega = 0.$$

We note first that the integral of ω taken over any circle parallel to the equator $\mathbf{c}^1$ vanishes. For example, if $0 < t_0 < 1$, then the 2-chain

$$\mathbf{c}^2 = [0, t_0] \times \mathbf{S}^1$$

has boundary

$$\partial \mathbf{c}^2 = \{t_0\} \times \mathbf{S}^1 - \mathbf{c}^1;$$

hence,

$$\int_{\{t_0\}\times \mathbf{S}^1} \omega = \int_{\{t_0\}\times \mathbf{S}^1} \omega - \int_{\mathbf{c}^1} \omega = \int_{\partial \mathbf{c}^2} \omega = \int_{\mathbf{c}^2} d\omega = \int_{\mathbf{c}^2} 0 = 0.$$

Having this, we consider the form $\phi^*\omega$, a 1-form on the infinite strip—a space which, from the viewpoint of its differentiable structure alone, is indistinguishable from $\mathbf{E}^2$. We have

$$d(\phi^*\omega) = \phi^*(d\omega) = \phi^*(0) = 0,$$

and hence $\phi^*\omega$ is a closed 1-form on the strip. By the converse of the Poincaré lemma, there exists a function g on the strip such that

$$\phi^*\omega = dg.$$

The crucial question follows: Is there a function f on the cylinder such that $\phi^*f = g$? Clearly, for this condition it is necessary and sufficient that g be periodic of period 2π in θ. But

$$g(t, \theta + 2\pi) - g(t, \theta) = \int_\theta^{\theta+2\pi} \left[\frac{d}{ds} g(t, s) \right] ds = \int_{\{t\}\times[\theta, \theta+2\pi]} dg$$

$$= \int_{\{t\}\times[\theta, \theta+2\pi]} \phi^*\omega = \int_{\{t\}\times \mathbf{S}^1} \omega = 0.$$

Thus, g has the required periodicity, so there is a function f on the cylinder satisfying $\phi^*f = g$. Hence, $dg = d(\phi^*f) = \phi^*(df)$, and

$$\phi^*\omega = \phi^*(df).$$

The covering mapping ϕ is locally one-one with a smooth inverse; hence ϕ^* is one-one and, finally,

$$\omega = df.$$

The reader should think about 2-forms on the cylinder.

3. Our next example is the 2-*sphere*, $\mathbf{S}^2$. We represent this explicitly in $\mathbf{E}^3$ as the locus of $x^2 + y^2 + z^2 = 1$. We cover $\mathbf{S}^2$ with the 2-coordinate neighborhoods,

$$\mathbf{U}^+ = \{z > -\tfrac{1}{2}\}, \qquad \mathbf{U}^- = \{z < \tfrac{1}{2}\}$$

which overlap in the equatorial zone,

$$\mathbf{U}^+ \cap \mathbf{U}^- = \{-\tfrac{1}{2} < z < \tfrac{1}{2}\}.$$

The set U^+ is diffeomorphic (that is, differentially isomorphic) to E^2, say, by projecting it onto the equatorial plane from the point $(0, 0, -\frac{1}{2})$; so is U^-. The set $U^+ \cap U^{-1}$ is diffeomorphic to the cylinder, a result of the Mercator projection. We shall first prove the following remark:

(6.12) PROPOSITION: *Each closed 1-form on S^2 is exact.*

Proof: Let α be a 1-form such that $d\alpha = 0$. By what we know for E^2, there exists a function g on U^+ such that $\alpha = dg$ on U^+ and a function h on U^- such that $\alpha = dh$ on U^-. Now,

$$d(g - h) = dg - dh = \alpha - \alpha = 0 \quad \text{on } U^+ \cap U^-.$$

Since the zone $U^+ \cap U^-$ is connected, we conclude that $g - h$ is constant on this set—that is, $g - h = c$ on $U^+ \cap U^-$.

We now change the function h on all U^- to the function $h_1 = h + c$. Then $dg = \alpha$ on U^+, $dh_1 = d(h + c) - dh = \alpha$ on U^-, and $g = h_1$ on $U^+ \cap U^-$. We define a function f on S^2 by setting $f = g$ on U^+ and $f = h_1$ on U^-—valid precisely because $g = h_1$ on the overlap—and have $\alpha = df$ as required.

For 2-forms we may write the following:

PROPOSITION: *A 2-form ω on S^2 is exact if and only if*

$$(6.13) \qquad\qquad \int_{S^2} \omega = 0.$$

Proof: This proof is more difficult. We give S^2 the orientation of the outward normal and give the equator,

$$c^1 = \{z = 0\},$$

the orientation of counterclockwise rotation in the (x, y)-plane, with the understanding that x, y, z is a right-handed system. Then,

$$c^1 = \partial H^+ = -\partial H^-,$$

where $H^+ = \{z > 0\}$ and $H^- = \{z < 0\}$ are the upper and lower hemispheres, respectively.

In one direction, the proof is easy. If $\omega = d\alpha$, then

$$\int_{S^2} \omega = \int_{S^2} d\alpha = \int_{\partial S^2} \alpha = 0.$$

Conversely, we assume ω is a 2-form on S^2 satisfying the relation,

(6.14) $$\int_{\mathbf{S}^2} \omega = 0.$$

Because $\mathbf{U}^+$ and $\mathbf{U}^-$ are diffeomorphic to $\mathbf{E}^2$, there is a 1-form β on $\mathbf{U}^+$ such that $\omega = d\beta$ on $\mathbf{U}^+$ and a 1-form γ on $\mathbf{U}^-$ such that $\omega = d\gamma$ on $\mathbf{U}^-$. The 1-form $\beta - \gamma$, defined on $\mathbf{U}^+ \cap \mathbf{U}^-$, is closed, because

$$d(\beta - \gamma) = d\beta - d\gamma = \omega - \omega = 0.$$

As a result of Equation (6.9), this form, $\beta - \gamma$, will be exact on the zone if its integral over $\mathbf{c}^1$ vanishes. Now

$$\int_{\mathbf{c}^1} \beta = \int_{\partial \mathbf{H}^+} \beta = \int_{\mathbf{H}^+} d\beta = \int_{\mathbf{H}^+} \omega,$$

$$\int_{\mathbf{c}^1} \gamma = -\int_{\partial \mathbf{H}^-} \gamma = -\int_{\mathbf{H}^-} d\gamma = -\int_{\mathbf{H}^-} \omega,$$

$$\int_{\mathbf{c}^1} (\beta - \gamma) = \int_{\mathbf{H}^+} \omega + \int_{\mathbf{H}^-} \omega = \int_{\mathbf{S}^2} \omega = 0$$

by (6.14). Hence, there is a function h on $\mathbf{U}^+ \cap \mathbf{U}^-$ such that

$$\beta - \gamma = dh \quad \text{on } \mathbf{U}^+ \cap \mathbf{U}^-.$$

We now encounter technical difficulties. If we could prolong the function h to a function defined on all of $\mathbf{U}^-$, then we would replace γ by $\gamma + dh$ and quickly end the argument. Unfortunately we cannot, because h may get nasty as we approach the lower boundary $\{z = -\frac{1}{2}\}$ of the equatorial zone. However, if we cut back the zone size, it is possible to prolong h from what remains.

To do so, we assume for the moment that we can find a (smooth) function g on $\mathbf{U}^-$ such that $g = 1$ on $\{0 < z < \frac{1}{2}\}$ and $g = 0$ on $\{z < -\frac{1}{4}\}$. We then define h_1 on $\mathbf{U}^-$ by

$$\left\{ \begin{array}{l} h_1 = gh \quad \text{on } \mathbf{U}^+ \cap \mathbf{U}^-, \\ h_1 = 0 \quad \text{on } \{z \leqq -\frac{1}{4}\}. \end{array} \right.$$

Finally, we define α on $\mathbf{S}^2$ by

$$\left\{ \begin{array}{l} \alpha = \beta \quad \text{on } \mathbf{H}^+, \\ \alpha = \gamma + dh_1 \quad \text{on } \mathbf{U}^-, \end{array} \right.$$

which is a definition because on the common part, $\mathbf{H}^+ \cap \mathbf{U}^-$, of these sets we have

$$\gamma + dh_1 = \gamma + d(gh) = \gamma + dh = \beta.$$

Next,

$$\begin{cases} d\alpha = d\beta = \omega & \text{on } \mathbf{H}^+, \\ d\alpha = d(\gamma - dh_1) = d\gamma = \omega & \text{on } \mathbf{H}^-. \end{cases}$$

Therefore, $d\alpha = \omega$ on $\mathbf{S}^2$.

For the function g, we may take a function of the height z alone. We thus seek a function $g = g(z) \in C^\infty$, which vanishes for $z < -\frac{1}{4}$ and equals 1 for $z > 0$. The standard method for construction of such a function proceeds in several steps which we shall outline briefly: (1) $A(z) = 0$ for $z \leq 0$, $A(z) = \exp(-1/z)$ for $z > 0$; (2) $B(z) = A(z) A(1 - z)$, so $B(z) > 0$ for $0 < z < 1$, $B(z) = 0$ otherwise; (3) $C(z) = \int_0^z B(t)\, dt \Big/ \int_0^1 B(t)\, dt$ so that $C(z) = 0$ for $z \leq 0$, $C(z) = 1$ for $z \geq 1$; and (4) $g(z) = C(4z + 1)$.

Keeping this example before you, you should be able to work out the de Rham groups for $\mathbf{S}^3$ and then for $\mathbf{S}^n$. [Note that the element of area on $\mathbf{S}^2$ is an example of a 2-form with nonzero integral.]

4. The next example we shall discuss is the *projective plane*, $\mathbf{P}^2$. The projective plane is obtained from $\mathbf{S}^2$ by identifying antipodal points. Thus, $\mathbf{S}^2$ is a 2-sheeted covering surface of $\mathbf{P}^2$; we let π denote the covering mapping. If α denotes the mapping

$$\alpha(\mathbf{x}) = -\mathbf{x}$$

on $\mathbf{S}^2$ which maps each point $\mathbf{x}$ of $\mathbf{S}^2$ to its antipode, then $\pi \circ \alpha = \pi$ —that is, the diagram

$$(6.15) \qquad \mathbf{S}^2 \xrightarrow{\ \alpha\ } \mathbf{S}^2$$
$$\pi \searrow \quad \swarrow \pi$$
$$\mathbf{P}^2$$

is commutative. If ω is a p-form on $\mathbf{P}^2$ and $\pi^*\omega = 0$, then $\omega = 0$. This statement is true because locally π is a diffeomorphism (indeed, this is the method by which we define the manifold structure on $\mathbf{P}^2$); hence π^* is an isomorphism. Our crucial tool for constructing forms on $\mathbf{P}^2$ follows.

PROPOSITION: *Let η be a p-form on $\mathbf{S}^2$ ($p = 0, 1, 2$). Then there is a p-form ω on $\mathbf{P}^2$ such that*

(6.16) $$\pi^*\omega = \eta,$$

if and only if

(6.17) $$\alpha^*\eta = \eta.$$

Proof: Suppose $\eta = \pi^*\omega$. Then

$$\alpha^*\eta = \alpha^*\pi^*\omega = (\pi \circ \alpha)^*\omega = \pi^*\omega = \eta.$$

Conversely, let η be a form on $\mathbf{S}^2$ which satisfies $\alpha^*\eta = \eta$. We select a covering of $\mathbf{S}^2$ by open sets $\mathbf{U}$ which are so small that $\alpha\mathbf{U} \cap \mathbf{U} = \varnothing$. The sets $\pi(\mathbf{U})$ then cover $\mathbf{P}^2$, and for each $\mathbf{U}$,

$$\pi : \mathbf{U} \rightarrow \pi(\mathbf{U})$$

is a diffeomorphism. Let $\lambda_\mathbf{U}$ denote its inverse:

$$\lambda_\mathbf{U} : \pi(\mathbf{U}) \rightarrow \mathbf{U}.$$

We define

$$\omega_{\pi(\mathbf{U})} = \lambda_\mathbf{U}^*(\eta) \quad \text{on } \pi(\mathbf{U})$$

so that

$$\pi^*(\omega_{\pi(\mathbf{U})}) = \pi^*\lambda_\mathbf{U}^*(\eta) = (\lambda_\mathbf{U} \circ \pi)^*\eta = (\text{identity})^*\eta = \eta \quad \text{on } \mathbf{U}.$$

We have defined a p-form on each open set $\pi(\mathbf{U})$ of the covering of $\mathbf{P}^2$. Do these fit together? Suppose $\pi(\mathbf{U}) \cap \pi(\mathbf{V}) \neq \varnothing$. If $\mathbf{U}$ meets $\mathbf{V}$, then

$$\pi(\mathbf{U} \cap \mathbf{V}) = \pi(\mathbf{U}) \cap \pi(\mathbf{V}),$$

$$\lambda_\mathbf{U} = \lambda_\mathbf{V} \quad \text{on } \pi(\mathbf{U}) \cap \pi(\mathbf{V}),$$

$$\omega_{\pi(\mathbf{U})} = \lambda_\mathbf{U}^*(\eta) = \lambda_\mathbf{V}^*(\eta) = \omega_{\pi(\mathbf{V})} \quad \text{on } \pi(\mathbf{U}) \cap \pi(\mathbf{V}).$$

Otherwise, $\mathbf{U}$ meets $\alpha\mathbf{V}$,

$$\pi(\mathbf{U} \cap \alpha\mathbf{V}) = \pi(\mathbf{U}) \cap \pi(\mathbf{V}),$$

$$\lambda_\mathbf{U} = \alpha \circ \lambda_\mathbf{V} \quad \text{on } \pi(\mathbf{U}) \cap \pi(\mathbf{V}),$$

$$\omega_{\pi(\mathbf{U})} = \lambda_\mathbf{U}^*(\eta) = \lambda_\mathbf{V}^*\alpha^*\eta = \lambda_\mathbf{V}^*(\eta) = \omega_{\pi(\mathbf{V})} \quad \text{on } \pi(\mathbf{U}) \cap \pi(\mathbf{V}),$$

which shows us that the forms we have defined do fit into a single form ω on $\mathbf{P}^2$ such that $\pi^*\omega = \eta$.

This proof may seem unduly fussy for something which one usually passes off as "obvious," an acceptable procedure only if one can do the details mentally, a feat usually learned by seeing such details spelled out once.

PROPOSITION: *If ω is a closed 1-form on $\mathbf{P}^2$, then ω is exact.*

Proof: The form $\pi^*\omega$ is a closed 1-form on $\mathbf{S}^2$. By (6.12), there is a function (0-form) f on $\mathbf{S}^2$ such that $\pi^*\omega = df$. We have

$$d(\alpha^*f) = \alpha^*(df) = \alpha^*\pi^*\omega = (\pi \circ \alpha)^*\omega = \pi^*\omega = df.$$

Hence,

$$d(\alpha^*f - f) = 0 \quad \text{on } \mathbf{S}^2.$$

It follows that $\alpha^*f - f = c$, a constant, on $\mathbf{S}^2$—that is,

$$f(-\mathbf{x}) - f(\mathbf{x}) = c \quad \text{for each } \mathbf{x} \in \mathbf{S}^2.$$

Apply this relation to $-\mathbf{x}$:

$$f(\mathbf{x}) - f(-\mathbf{x}) = c;$$

we deduce $c = 0$,

$$\alpha^*f = f.$$

By (6.16), there is a function g on $\mathbf{P}^2$ satisfying $\pi^*g = f$; hence,

$$\pi^*dg = d\pi^*g = df = \pi^*\omega, \qquad dg = \omega.$$

This result together with the following proposition show that the de Rham groups of $\mathbf{P}^2$ are trivial for $p = 1, 2$.

PROPOSITION: *Each 2-form on $\mathbf{P}^2$ is exact.*

(Note that no integral condition is required as in (6.13), because the surface $\mathbf{P}^2$ is nonorientable and consequently has no two-dimensional real homology.)

Proof: Let σ be a 2-form on $\mathbf{P}^2$. Then $\tau = \pi^*\sigma$ is a 2-form on $\mathbf{S}^2$ which satisfies $\alpha^*\tau = \tau$. Hence,

$$\int_{\mathbf{S}^2} \tau = \int_{\mathbf{S}^2} \alpha^*\tau = \int_{\alpha_*\mathbf{S}^2} \tau = \int_{-\mathbf{S}^2} \tau = -\int_{\mathbf{S}^2} \tau, \qquad \int_{\mathbf{S}^2} \tau = 0.$$

[The mapping $(x^1, x^2, x^3) \rightarrow (-x^1, -x^2, -x^3)$ reverses the orientation on $\mathbf{S}^2$.] We now may apply (6.13): There is a 1-form η on $\mathbf{S}^2$ such that $\tau = d\eta$. Now

$$d(\alpha^*\eta) = \alpha^*d\eta = \alpha^*\tau = \alpha^*(\pi^*\sigma) = (\pi \circ \alpha)^*\sigma = \pi^*\sigma = \tau = d\eta.$$

Hence,

$$d(\alpha^*\eta - \eta) = 0 \quad \text{on } \mathbf{S}^2.$$

By (6.12), there is a function f on $\mathbf{S}^2$ satisfying the equation,

$$\alpha^*\eta - \eta = df.$$

We apply α^* to this relation to find

$$\eta - \alpha^*\eta = \alpha^* df.$$

Hence,

$$\alpha^* df = -df.$$

We now set

$$\eta_1 = \eta + \tfrac{1}{2}df.$$

Then,

$$d\eta_1 = d\eta = \pi^*\sigma,$$

and

$$\alpha^*\eta_1 = \alpha^*\eta + \tfrac{1}{2}\alpha^* df = \eta + df - \tfrac{1}{2}df = \eta_1.$$

By (6.16), there is a 1-form ω on $\mathbf{P}^2$ such that $\pi^*\omega = \eta_1$. Hence,

$$\pi^* d\omega = d\pi^*\omega = d\eta_1 = \pi^*\sigma, \qquad d\omega = \sigma.$$

See what you can do with $\mathbf{P}^3$ and $\mathbf{P}^n$. Another interesting exercise is to work out the situation first for the ordinary torus, $\mathbf{T}^2 = \mathbf{S}^1 \times \mathbf{S}^1$, and then for the n-dimensional torus.

7. AN EXAMPLE OF A MOVING FRAME

Suppose we have vectors $\mathbf{v}_1, \cdots, \mathbf{v}_n$ in $\mathbf{E}^n$, each of which is a smooth function $\mathbf{v}_i = \mathbf{v}_i(u^1, \cdots, u^r)$ of $(u^1, \cdots, u^r)$, which varies over a domain in r-space. We assume that for each point in u-space, the vectors $\mathbf{v}_1, \cdots, \mathbf{v}_n$ are linearly independent in $\mathbf{E}^n$ and consequently form a linear basis of $\mathbf{E}^n$. If we let $\mathbf{e}_1, \cdots, \mathbf{e}_n$ denote the standard basis of $\mathbf{E}^n$ [$\mathbf{e}_1 = (1, 0, \cdots, 0)$, $\mathbf{e}_2 = (0, 1, 0, \cdots, 0)$, etc.] then we may write

$$(7.1) \qquad \mathbf{v}_i = (a_i^1, a_i^2, \cdots, a_i^n) = \sum a_i^j \mathbf{e}_j.$$

The $a_i^j = a_i^j(u^1, \cdots, u^n)$ make up an $n \times n$ matrix,

$$(7.2) \qquad A = \|a_i^j\|,$$

which is nonsingular at each point of the u-space by our assumption that $v_1, \cdots, v_n$ are linearly independent.

We differentiate (7.1) as follows:

$$(7.3) \quad d\mathbf{v}_i = (da_i^1, \cdots, da_i^n) = \Sigma (da_i^j)\mathbf{e}_j = \Sigma \left(\frac{\partial a_i^j}{\partial u^k}\right)(du^k)\mathbf{e}_j.$$

These equations give us the components of the differential of each $\mathbf{v}_i$ in terms of the standard basis $\mathbf{e}_1, \cdots, \mathbf{e}_n$. The moving frame idea, as applied to this situation, is to express these differentials of the vectors $\mathbf{v}_i$ in terms of the $\mathbf{v}_1, \cdots, \mathbf{v}_n$ themselves. We accomplish this by first inverting the equations (7.1) (possible because A is nonsingular) to express the $\mathbf{e}_i$ as linear combinations of the $\mathbf{v}_j$, and then substituting these expressions in Equation (7.3). What results is

$$(7.4) \qquad d\mathbf{v}_i = \Sigma \omega_i^j \mathbf{v}_j$$

where the ω_i^j are 1-forms.

Precisely, if we write

$$(7.5) \qquad \mathbf{v} = \begin{pmatrix} \mathbf{v}_1 \\ \cdot \\ \cdot \\ \cdot \\ \mathbf{v}_n \end{pmatrix}, \qquad \mathbf{e} = \begin{pmatrix} \mathbf{e}_1 \\ \cdot \\ \cdot \\ \cdot \\ \mathbf{e}_n \end{pmatrix}, \qquad \Omega = \|\omega_i^j\|,$$

then

$$\mathbf{v} = A\mathbf{e}, \qquad \mathbf{e} = A^{-1}\mathbf{v}, \qquad d\mathbf{v} = (dA)\mathbf{e} = (dA)A^{-1}\mathbf{v},$$
$$(7.6) \qquad\qquad \Omega = (dA)A^{-1}.$$

As an application, we pose the following question: What is the differential of $(\det A)$? We recall that we may consider the determinant of a matrix to be an alternating multilinear functional of the rows of the matrix,

$$(7.7) \qquad \det A = \Delta(\mathbf{v}_1, \cdots, \mathbf{v}_n).$$

Therefore, the rule for differentiating a product is applicable; for example,

$$\frac{\partial}{\partial u^k}(\det A) = \Delta\left(\frac{\partial \mathbf{v}_1}{\partial u^k}, \mathbf{v}_2, \cdots, \mathbf{v}_n\right) + \Delta\left(\mathbf{v}_1, \frac{\partial \mathbf{v}_2}{\partial u^k}, \mathbf{v}_3, \cdots, \mathbf{v}_n\right)$$
$$+ \cdots + \Delta\left(\mathbf{v}_1, \cdots, \mathbf{v}_{n-1}, \frac{\partial \mathbf{v}_n}{\partial u^k}\right).$$

Multiplying by du^k and summing, we have

(7.8) $d(\det A) = \Delta(d\mathbf{v}_1, \mathbf{v}_2, \cdots, \mathbf{v}_n) + \Delta(\mathbf{v}_1, d\mathbf{v}_2, \mathbf{v}_3, \cdots, \mathbf{v}_n)$
$$+ \cdots + \Delta(\mathbf{v}_1, \cdots, \mathbf{v}_{n-1}, d\mathbf{v}_n).$$

Now,

$$\Delta(d\mathbf{v}_1, \mathbf{v}_2, \cdots, \mathbf{v}_n) = \Delta(\textstyle\sum \omega_1^j \mathbf{v}_j, \mathbf{v}_2, \cdots, \mathbf{v}_n)$$
$$= \textstyle\sum \omega_1^j \Delta(\mathbf{v}_j, \mathbf{v}_2, \cdots, \mathbf{v}_n)$$
$$= \omega_1^1 \Delta(\mathbf{v}_1, \cdots, \mathbf{v}_n)$$
$$= \omega_1^1 (\det A),$$

because determinants with two equal rows vanish. We come to

(7.9) $d(\det A) = \omega_1^1(\det A) + \omega_2^2(\det A) + \cdots + \omega_n^n(\det A)$
$$= (\omega_1^1 + \omega_2^2 + \cdots + \omega_n^n)(\det A).$$

In view of Equation (7.6), we may write this relation as

(7.10) $$\frac{d(\det A)}{(\det A)} = \text{trace}[(dA) A^{-1}].$$

This formula is useful in differential geometry.

8. THE GAUSS-BONNET THEOREM FOR SURFACES

In this section, $\mathbf{M}$ is a compact oriented manifold of dimension two which has a Riemannian structure. This last condition means that each tangent plane at each point of $\mathbf{M}$ has a euclidean structure—inner products are defined—and if $\mathbf{v}$ and $\mathbf{w}$ are smooth vector fields over a region on $\mathbf{M}$, then $\mathbf{v} \cdot \mathbf{w}$ is a smooth scalar over that region. (The case to keep in mind is the classical one in which $\mathbf{M}$ is a closed surface in $\mathbf{E}^3$. The tangent planes at the various points of $\mathbf{M}$ inherit their euclidean structures from that of the ambient space, $\mathbf{E}^3$. From topology, we know that a closed surface in $\mathbf{E}^3$ divides $\mathbf{E}^3$ into inside and outside regions. Taking the outward normal imposes an orientation on $\mathbf{M}$ by the right-hand rule.)

Over a local coordinate neighborhood $\mathbf{U}$ on $\mathbf{M}$, we may find vector fields $\mathbf{e}_1$ and $\mathbf{e}_2$, which make up a right-handed orthonormal frame in the tangent space of each point of **U**. Letting σ_1 and σ_2 be the dual basis of 1-forms,† we may write, symbolically,

$$(8.1) \qquad d\mathbf{x} = \sigma_1 \mathbf{e}_1 + \sigma_2 \mathbf{e}_2.$$

In the case of an imbedded surface in $\mathbf{E}^3$ this relation is precise. On an abstract surface, we use it in order to motivate our steps—in particular, to "see" the transformation rule for the σ's. We may consider Equation (8.1) to mean that an infinitesimal displacement of the point **x** on the surface will have components σ_1 in the $\mathbf{e}_1$ direction and σ_2 in the $\mathbf{e}_2$ direction.

We may compute σ_1 and σ_2 explicitly as follows. We start with a local coordinate system, u^1 and u^2. Then

$$\frac{\partial}{\partial u^i} \cdot \frac{\partial}{\partial u^j} = g_{ij}(u^1, u^2)$$

gives us a positive definite matrix $\|g_{ij}\|$. Indeed, the Riemannian structure on **M** is usually given in advance by means of such matrices, one for each chart of a covering with appropriate relations on overlapping charts. By some process of orthonormalization we construct the orthonormal frame $\mathbf{e}_1$ and $\mathbf{e}_2$, and we have

$$(8.2) \qquad \begin{cases} \mathbf{e}_1 = a_{11} \dfrac{\partial}{\partial u^1} + a_{12} \dfrac{\partial}{\partial u^2} \\[2mm] \mathbf{e}_2 = a_{21} \dfrac{\partial}{\partial u^1} + a_{22} \dfrac{\partial}{\partial u^2}. \end{cases}$$

Substitution of this quantity into

$$d\mathbf{x} = \sigma_1 \mathbf{e}_1 + \sigma_2 \mathbf{e}_2 = du^1 \frac{\partial}{\partial u^1} + du^2 \frac{\partial}{\partial u^2}$$

---

† The duality between vector fields and 1-forms is most simply explained by the natural bases associated with a local coordinate system $u^1, \cdots, u^n$ (in arbitrary dimension). Here, $\mathbf{e}_1 = \partial/\partial u^1, \cdots, \mathbf{e}_n = \partial/\partial u^n$ is a basis of the tangent space, and $du^1, \cdots, du^n$ is a basis of the form space. The dual pairing between the two is

$$\left(du^i, \frac{\partial}{\partial u^j} \right) = \delta^i_j,$$

which is easily seen to be independent of local coordinates. Details are found in Sternberg, p. 72, or Kobayashi and Nomizu, p. 6.

gives us

$$(8.3) \qquad \begin{cases} a_{11}\sigma_1 + a_{21}\sigma_2 = du^1 \\ a_{12}\sigma_1 + a_{22}\sigma_2 = du^2, \end{cases}$$

which we may solve for σ_1 and σ_2. We note that the relations $\mathbf{e}_i \cdot \mathbf{e}_j = \delta_{ij}$ imply that

$$(8.4) \qquad \delta_{ij} = \sum a_{ik}g_{kl}a_{jl}.$$

The element of area on $\mathbf{M}$ is the 2-form

$$(8.5) \qquad \sigma_1\sigma_2 = \sqrt{g}\, du^1\, du^2$$

where

$$(8.6) \qquad g = \begin{vmatrix} g_{11} & g_{12} \\ g_{21} & g_{22} \end{vmatrix}.$$

This relation follows from (8.3), which implies $[\det (a_{ij})]\sigma_1\sigma_2 = du^1\, du^2$, and from (8.4), which implies $[\det (a_{ij})]^2 g = 1$. (Note that $\det (a_{ij}) > 0$, because both frames agree with the orientation.)

We next claim there exists a unique 1-form ϖ on our neighborhood $\mathbf{U}$, such that

$$(8.7) \qquad \begin{cases} d\sigma_1 = \varpi\sigma_2 \\ d\sigma_2 = -\varpi\sigma_1, \end{cases}$$

which is true because $\sigma_1\sigma_2$ is a basis of 2-forms. Hence, $d\sigma_1 = a_1\sigma_1\sigma_2$, $d\sigma_2 = a_2\sigma_1\sigma_2$, and we are forced to the unique solution $\varpi = a_1\sigma_1 + a_2\sigma_2$.

Suppose we have a second coordinate neighborhood $\overline{\mathbf{U}}$ which overlaps $\mathbf{U}$, and an orthonormal frame $\bar{\mathbf{e}}_1$, $\bar{\mathbf{e}}_2$ for $\overline{\mathbf{U}}$. Then,

$$(8.8) \qquad d\mathbf{x} = \bar{\sigma}_1\bar{\mathbf{e}}_1 + \bar{\sigma}_2\bar{\mathbf{e}}_2$$

on $\overline{\mathbf{U}}$. On the intersection $\mathbf{U} \cap \overline{\mathbf{U}}$, we have

$$(8.9) \qquad \begin{cases} \bar{\mathbf{e}}_1 = (\cos \alpha)\mathbf{e}_1 + (\sin \alpha)\mathbf{e}_2 \\ \bar{\mathbf{e}}_2 = -(\sin \alpha)\mathbf{e}_1 + (\cos \alpha)\mathbf{e}_2, \end{cases}$$

because $\mathbf{e}_1$, $\mathbf{e}_2$ and $\bar{\mathbf{e}}_1$, $\bar{\mathbf{e}}_2$ are both right-handed orthonormal systems. Here α is a scalar on $\mathbf{U} \cap \overline{\mathbf{U}}$. Because (8.1) and (8.8) are both valid on $\mathbf{U} \cap \overline{\mathbf{U}}$, we have

(8.10)
$$\begin{cases} \sigma_1 = (\cos \alpha)\bar{\sigma}_1 - (\sin \alpha)\bar{\sigma}_2 \\ \sigma_2 = (\sin \alpha)\bar{\sigma}_1 + (\cos \alpha)\bar{\sigma}_2. \end{cases}$$

On $\bar{\mathbf{U}}$, we have

(8.11)
$$\begin{cases} d\bar{\sigma}_1 = \bar{\varpi}\bar{\sigma}_2 \\ d\bar{\sigma}_2 = -\bar{\varpi}\bar{\sigma}_1. \end{cases}$$

To find the relation between ϖ and $\bar{\varpi}$ on $\mathbf{U} \cap \bar{\mathbf{U}}$, we differentiate the relations in (8.10):

$$\begin{aligned} d\sigma_1 &= -(\sin \alpha)(d\alpha)\bar{\sigma}_1 - (\cos \alpha)(d\alpha)\bar{\sigma}_2 \\ &\quad + (\cos \alpha)\bar{\varpi}\bar{\sigma}_2 + (\sin \alpha)\bar{\varpi}\bar{\sigma}_1 \\ &= (\bar{\varpi} - d\alpha)\bar{\sigma}_2; \\ d\sigma_2 &= (\cos \alpha)(d\alpha)\bar{\sigma}_1 - (\sin \alpha)(d\alpha)\bar{\sigma}_2 \\ &\quad + (\sin \alpha)\bar{\varpi}\bar{\sigma}_2 - (\cos \alpha)\bar{\varpi}\bar{\sigma}_1 \\ &= -(\bar{\varpi} - d\alpha)\bar{\sigma}_1. \end{aligned}$$

It follows that

(8.12)
$$\varpi = \bar{\varpi} - d\alpha.$$

The important consequence is that the 2-form $d\varpi = d\bar{\varpi}$ is defined on all of $\mathbf{M}$, independent of the moving frame. We set

(8.13)
$$d\varpi + K\sigma_1\sigma_2 = 0,$$

defining the *Gaussian* or *total curvature* K of $\mathbf{M}$. We shall prove the Gauss-Bonnet theorem:

(8.14)
$$\int_{\mathbf{M}} K\sigma_1\sigma_2 = 2\pi\chi(\mathbf{M})$$

where $\chi(\mathbf{M})$ is the *Euler-Poincaré characteristic* of $\mathbf{M}$.

We begin by introducing a new space—the space $\mathbf{B}$ of all unit tangents $\mathbf{v}$ at all points of $\mathbf{M}$. This space is three-dimensional, and there exists a natural mapping (projection)

(8.15)
$$p : \mathbf{B} \to \mathbf{M}$$

which sends each tangent vector $\mathbf{v}$ to the (base) point $p(\mathbf{v})$ at which it is a tangent vector. If $\mathbf{U}$ is the preceding local coordinate neighborhood, and $\mathbf{x}$ is any point of $\mathbf{U}$, the typical unit tangent vector at $\mathbf{x}$ is

(8.16) $\mathbf{v} = (\cos \phi)\mathbf{e}_1|_\mathbf{x} + (\sin \phi)\mathbf{e}_2|_\mathbf{x}.$

Since $\mathbf{U}$ is parameterized by u^1, u^2, we see that

$$u^1, u^2, \phi$$

parameterize the set $p^{-1}(\mathbf{U})$ in $\mathbf{B}$. In this way, we turn $\mathbf{B}$ into a differentiable manifold. (You should recognize $\mathbf{B}$ as a fiber bundle with $\mathbf{M}$ as the base space, p as projection, the fiber as the unit circle, and the group as the rotation group.) The differential forms,

$$du^1, du^2, d\phi,$$

or, equivalently,

(8.17) $\sigma_1, \sigma_2, d\phi,$

form a basis for 1-forms on that part $p^{-1}(\mathbf{U})$ of $\mathbf{B}$ over $\mathbf{U}$.

We can, and should, be more precise. The forms σ_1, σ_2 are defined on $\mathbf{U}$, and we have

$$p : p^{-1}(\mathbf{U}) \to \mathbf{U}.$$

Thus, what we are really discussing are

$$p^*\sigma_1, p^*\sigma_2, d\phi,$$

which form a basis of 1-forms on $p^{-1}(\mathbf{U})$.

If we consider what happens on the common part $\mathbf{U} \cap \bar{\mathbf{U}}$ of two coordinate neighborhoods, we have, using the notation of (8.8), etc.,

$$\mathbf{v} = (\cos \bar{\phi})\bar{\mathbf{e}}_1 + (\sin \bar{\phi})\bar{\mathbf{e}}_2$$

$$= (\cos \bar{\phi} \cos \alpha - \sin \bar{\phi} \sin \alpha)\mathbf{e}_1 + (\cos \bar{\phi} \sin \alpha + \sin \bar{\phi} \cos \alpha)\mathbf{e}_2$$

$$= [\cos (\bar{\phi} + \alpha)]\mathbf{e}_1 + [\sin (\bar{\phi} + \alpha)]\mathbf{e}_2.$$

Hence, comparison with (8.16) yields

(8.18) $\phi = \bar{\phi} + \alpha.$

We shall now define three 1-forms on $\mathbf{B}$; here we mean all of $\mathbf{B}$, not just a neighborhood. Each of these forms, which we call Σ_1, Σ_2, and Π, respectively, we first define on each $p^{-1}(\mathbf{U})$. Then we show that the "pieces of a form" coincide on the intersections $p^{-1}(\mathbf{U})$ $\cap p^{-1}(\bar{\mathbf{U}}) = p^{-1}(\mathbf{U} \cap \bar{\mathbf{U}})$.

First we define Σ_1 and Σ_2 on $p^{-1}(\mathbf{U})$ by the equations

$$(8.19) \qquad \begin{cases} p^*\sigma_1 = (\cos \phi)\Sigma_1 - (\sin \phi)\Sigma_2 \\ p^*\sigma_2 = (\sin \phi)\Sigma_1 + (\cos \phi)\Sigma_2. \end{cases}$$

On $\bar{\mathbf{U}}$, the corresponding forms $\bar{\Sigma}_1, \bar{\Sigma}_2$ are defined by the analogous equations,

$$(8.19') \qquad \begin{cases} p^*\bar{\sigma}_1 = (\cos \bar{\phi})\bar{\Sigma}_1 - (\sin \bar{\phi})\bar{\Sigma}_2 \\ p^*\bar{\sigma}_2 = (\sin \bar{\phi})\bar{\Sigma}_1 + (\cos \bar{\phi})\bar{\Sigma}_2. \end{cases}$$

We consider the overlap $p^{-1}(\mathbf{U}) \cap p^{-1}(\bar{\mathbf{U}}) = p^{-1}(\mathbf{U} \cap \bar{\mathbf{U}})$, where (8.10) and (8.18) connect the various quantities [first applying p^* to (8.10)]. We wish to conclude that on this overlap, $\bar{\Sigma}_1 = \Sigma_1$ and $\bar{\Sigma}_2 = \Sigma_2$. We accomplish this most easily by expressing (8.19), (8.19'), and (8.10) with p^* applied in matrix notation:

$$(8.19) \qquad (p^*\sigma_1, p^*\sigma_2) = (\Sigma_1, \Sigma_2) \begin{pmatrix} \cos \phi & \sin \phi \\ -\sin \phi & \cos \phi \end{pmatrix};$$

$$(8.19') \qquad (p^*\bar{\sigma}_1, p^*\bar{\sigma}_2) = (\bar{\Sigma}_1, \bar{\Sigma}_2) \begin{pmatrix} \cos \bar{\phi} & \sin \bar{\phi} \\ -\sin \bar{\phi} & \cos \bar{\phi} \end{pmatrix};$$

$$(8.10') \qquad (p^*\sigma_1, p^*\sigma_2) = (p^*\bar{\sigma}_1, p^*\bar{\sigma}_2) \begin{pmatrix} \cos \alpha & \sin \alpha \\ -\sin \alpha & \cos \alpha \end{pmatrix}.$$

From these equations, we readily deduce

$$(\bar{\Sigma}_1, \bar{\Sigma}_2) \begin{pmatrix} \cos \bar{\phi} & \sin \bar{\phi} \\ -\sin \bar{\phi} & \cos \bar{\phi} \end{pmatrix} \begin{pmatrix} \cos \alpha & \sin \alpha \\ -\sin \alpha & \cos \alpha \end{pmatrix}$$
$$= (\Sigma_1, \Sigma_2) \begin{pmatrix} \cos \phi & \sin \phi \\ -\sin \phi & \cos \phi \end{pmatrix}.$$

But relation (8.18) tells us precisely that the product of the two rotation matrices on the left-hand side equals the rotation matrix on the right-hand side; therefore,

$$(\bar{\Sigma}_1, \bar{\Sigma}_2) = (\Sigma_1, \Sigma_2), \qquad \bar{\Sigma}_1 = \Sigma_1, \qquad \bar{\Sigma}_2 = \Sigma_2.$$

Thus, we have defined 1-forms Σ_1, Σ_2 on all of $\mathbf{B}$. The remaining form, Π, is defined on $p^{-1}(\mathbf{U})$ by

$$(8.20) \qquad p^*\varpi = \Pi - d\phi.$$

On $p^{-1}(\bar{\mathbf{U}})$, the analogous form is defined by

$$(8.20') \qquad p^*\bar{\varpi} = \bar{\varpi} - d\bar{\phi}.$$

But (8.12) gives us

$$(8.12') \qquad\qquad p^*\varpi = p^*\overline{\varpi} - p^* \, d\alpha$$

on $p^{-1}(\mathbf{U} \cap \overline{\mathbf{U}})$,† and (8.18) gives us

$$(8.18') \qquad\qquad d\phi = d\overline{\phi} + p^* \, d\alpha.$$

Combining these last four equations gives us $\Pi = \overline{\Pi}$. We have therefore succeeded in defining our third 1-form, Π, on all of $\mathbf{B}$.

From (8.19) we have (all exterior products)

$$(8.21) \qquad\qquad \Sigma_1\Sigma_2 = p^*(\sigma_1\sigma_2) = (p^*\sigma_1)(p^*\sigma_2).$$

From this equation and from (8.20),

$$(8.22) \qquad\qquad \Sigma_1\Sigma_2\Pi = (p^*\sigma_1)(p^*\sigma_2) \, d\phi,$$

because

$$(p^*\sigma_1)(p^*\sigma_2)(p^*\varpi) = p^*(\sigma_1\sigma_2\varpi) = 0.$$

From this equation, we deduce that Σ_1, Σ_2, and Π form a basis of the space of 1-forms on $\mathbf{B}$. Hence, any 2-form on $\mathbf{B}$ is a linear combination of $\Sigma_1\Sigma_2$, $\Pi\Sigma_1$, and $\Pi\Sigma_2$. We may work it out for $d\Sigma_1$, $d\Sigma_2$, and $d\Pi$. It suffices to work over one neighborhood of $p^{-1}(\mathbf{U})$. We differentiate the first formula in (8.19) and use (8.7) and (8.20) as needed:

$$(p^*\varpi)(p^*\sigma_2) = (d\phi)[-(\sin\phi)\Sigma_1 - (\cos\phi)\Sigma_2]$$
$$+ (\cos\phi)(d\Sigma_1) - (\sin\phi)(d\Sigma_2),$$

that is,

$$(\cos\phi)(d\Sigma_1) - (\sin\phi)(d\Sigma_2) = (\sin\phi)(\Pi\Sigma_1) + (\cos\phi)(\Pi\Sigma_2).$$

Similarly,

$$-(\sin\phi)(d\Sigma_1) - (\cos\phi)(d\Sigma_2) = (\cos\phi)(\Pi\Sigma_1) - (\sin\phi)(\Pi\Sigma_2).$$

Hence,

$$(8.23) \qquad\qquad \begin{cases} d\Sigma_1 = \Pi\Sigma_2 \\ d\Sigma_2 = -\Pi\Sigma_1. \end{cases}$$

---

† Note the slight inaccuracy in notation. We have identified the function (scalar) $p^*\alpha$ on $p^{-1}(\mathbf{U})$ with the scalar α on $\mathbf{U}$, just as we call the function x on the x-axis and the x-coordinate of the point (x, y) by the same letter.

Next, from (8.20), (8.13), and (8.21), we have

$$d\Pi = p^*(d\varpi) = -K(p^*\sigma_1)(p^*\sigma_2) = -K\Sigma_1\Sigma_2,$$

(8.24) $$d\Pi + K\Sigma_1\Sigma_2 = 0.$$

To illustrate the applicability of this relation, we begin with a special case of the Gauss-Bonnet theorem.

THEOREM: *If there exists a smooth unit tangent vector field on* **M**, *then*

(8.25) $$\int_{\mathbf{M}} K\sigma_1\sigma_2 = 0.$$

Proof: A unit field means a smooth mapping r which assigns to each point **x** of **M** a tangent vector at **x**. In other words,

(8.26) $$\begin{cases} r:\mathbf{M} \to \mathbf{B} \\ p \circ r = i = (\text{identity on } \mathbf{M}). \end{cases}$$

Since $\Sigma_1\Sigma_2 = p^*(\sigma_1\sigma_2)$, we have

$$r^*(\Sigma_1\Sigma_2) = (r^* \circ p^*)(\sigma_1\sigma_2) = (p \circ r)^*\sigma_1\sigma_2 = \sigma_1\sigma_2.$$

Hence, from (8.24),

$$d[r^*(\Pi)] + K\sigma_1\sigma_2 = 0.$$

The form $r^*\Pi$ is a 1-form on all of **M**. Therefore,

$$\int_{\mathbf{M}} K\sigma_1\sigma_2 = -\int_{\mathbf{M}} d(r^*\Pi) = -\int_{\partial(\mathbf{M})} r^*\Pi = 0,$$

because **M** is a compact surface and hence has no boundary.

For a surface to have a smooth vector field it is necessary and sufficient that its Euler characteristic vanish. This results from the following more general remarks.

If **M** is a compact oriented surface of Euler characteristic χ, there exists a vector field on **M** with a finite number of points deleted. The sum of the indices of the field at these singular points is exactly χ. It is even possible to find a field with only one singularity, and, if $\chi = 0$, with no singularities. These are standard facts in surface topology which we shall take for granted.

Suppose, then, we have a smooth unit tangent vector field defined on

$$\mathbf{M'} = \mathbf{M} - \{\mathbf{x}_1, \cdots, \mathbf{x}_n\},$$

which means we have a smooth mapping

$$r : \mathbf{M'} \to \mathbf{B}$$

such that

$$p \circ r = \text{(identity on } \mathbf{M'}).$$

As before, we now have

$$d(r^*\Pi) + K\sigma_1\sigma_2 = 0 \quad \text{on } \mathbf{M'}.$$

We take a small local coordinate neighborhood $\mathbf{U}_i$ centered at $\mathbf{x}_i$— so small that none of these $\mathbf{U}_i$ intersect and each $\mathbf{U}_i \cup \partial\mathbf{U}_i$ is diffeomorphic to the closed unit ball. Then

$$(8.27) \qquad \int_{\mathbf{M} - \cup_1^n \mathbf{U}_i} K\sigma_1\sigma_2 = -\int_{\mathbf{M} - \cup_1^n \mathbf{U}_i} d(r^*\Pi) = -\int_{\partial(\mathbf{M} - \cup_1^n \mathbf{U}_i)} r^*\Pi$$

$$= \sum_{i=1}^n \int_{\partial\mathbf{U}_i} r^*\Pi.$$

Let us examine one of these summands,

$$\int_{\partial\mathbf{U}_1} r^*\Pi.$$

We take a moving frame, $\mathbf{e}_1, \mathbf{e}_2$ on $\mathbf{U}_1$, so that (8.20) applies on $p^{-1}(\mathbf{U}_1)$,

$$\Pi = p^*\varpi + d\phi,$$

and so

$$r^*\Pi = \varpi + r^*\,d\phi \quad \text{on } \mathbf{U}_1 - \{\mathbf{x}_1\}.$$

Now

$$(8.28) \qquad \int_{\partial\mathbf{U}_1} \varpi = \int_{\mathbf{U}_1} d\varpi = -\int_{\mathbf{U}_1} K\sigma_1\sigma_2,$$

so that it remains to evaluate

$$\int_{\partial\mathbf{U}_1} r^*\,d\phi.$$

We recall the meaning of ϕ given by (8.16). At each point $\mathbf{x}$ of $\mathbf{U}_1 - \{\mathbf{x}_1\}$,

$$r(\mathbf{x}) = [\cos r^*(\phi)]\mathbf{e}_1 + [\sin r^*(\phi)]\mathbf{e}_2.$$

The angle $r^*(\phi) = \phi \circ r$ is only determined mod 2π, but its differential $d[r^*\phi] = r^* d\phi$ is completely determined. The boundary ∂U_1 is like a circle; it is a simple closed curve extending once around x_1. The integral of $r^* d\phi = d(r^*\phi)$ taken along it gives the variation of the angle $r^*\phi$ in following $r(x)$ over one circuit of this curve—that is,

$$(8.29) \qquad \int_{\partial U_1} r^* d\phi = 2\pi[\mathrm{ind}_{x_1}(r)].$$

Combining (8.27), (8.28), and (8.29), we arrive at the Gauss-Bonnet theorem,

$$(8.14) \qquad \int_M K\sigma_1\sigma_2 = 2\pi \sum_1^n \mathrm{ind}_{x_i}(r) = 2\pi\chi,$$

which proves that any two vector fields on M, each with a finite number of singularities, have the same index sum.

ON CONJUGATE AND CUT LOCI

Shoshichi Kobayashi

Let us fix a point p of a complete Riemannian manifold M and a geodesic g starting at p. Then a cut point of p along g is the first point q on g such that, for any point r on g beyond q, there is a geodesic from p to r shorter than g. In other words, q is the first point where g ceases to minimize. On the other hand, a point q is a conjugate point of p along g if there exists a 1-parameter family of geodesics from p to q neighboring g. (For a more precise definition, see Section 2.) If r is a point on g which lies beyond a conjugate point q of p, then g is not a minimizing geodesic between p and r.

In most books on classical differential geometry, the notion of cut point is rarely mentioned. What is worse, it is sometimes confused with that of conjugate point. The purpose of this article is to fill this gap by giving a self-contained account of basic properties of both cut points and conjugate points. The earliest paper on cut points was written by Poincaré (17). Although the subject was taken up again in the mid 1930's by Myers (15 and 16), and by J. H. C. Whitehead (28), the recent interest in cut points results from Klingenberg's papers (8 and 9) on Riemannian manifolds with positive curvature. Materials for the present article have been taken largely from Schoenberg (21), Synge (23), Preismann (18), Rauch (19), lecture notes by Ambrose and Singer at MIT, and the forthcoming Volume 2 of the book by Kobayashi and Nomizu (as well as from the papers previously mentioned). For better understanding, rigorous proofs are occasionally preceded by sketchy but intuitive proofs. When rigorous proofs require large machineries or extensive knowledge, less satisfactory but more elementary proofs are given instead. In the last section, we offer suggestions for further reading.

1. CUT LOCI

If we take two points, p and q, of a (connected) Riemannian manifold M and join them by a (continuous) piecewise differentiable curve, then we can measure the arc length of this curve using the Riemannian metric (often called the element of arc in older books). We consider all possible piecewise differentiable curves joining p and q and define the distance $d(p, q)$ between p and q as the infimum of their arc lengths. Then the distance function d satisfies the usual three axioms of a metric space, and we can talk about Cauchy sequences of points of M and also the completeness of M.

Hereafter, we shall assume that M is a complete Riemannian manifold. Since a compact metric space is complete, a compact Riemannian manifold is always complete. There are a few basic facts about geodesics of a complete Riemannian manifold which we must use in this article. If we say "a geodesic $g(t)$," we understand that it is parametrized by arc length and t is a parameter. A geodesic $g(t)$, which is originally defined only for some interval $a \leqq t \leqq b$, is said to be infinitely extendable if it can be extended to a geodesic $g(t)$ defined for the whole interval, $-\infty < t < \infty$. If we take as an example the $(x\text{-}y)$-plane with the usual metric and delete the origin, we then obtain an incomplete Riemannian manifold and the positive x-axis is not infinitely extendable.

The first basic fact we need is that, on a complete Riemannian manifold, every geodesic is infinitely extendable. This fact is intuitively clear and not difficult to prove. Although its converse is also true, its proof is much harder and we do not need it here. The second basic fact we need is that, on a complete Riemannian manifold, any two points can be joined by a minimizing geodesic. [A geodesic from p to q is said to be minimizing if its arc length gives the distance $d(p, q)$.] Two proofs of this fact are known, neither of which is very easy. The first is the work of Hilbert (6). [See also Hopf and Rinow (7) and Cartan (4).] The second proof is that of de Rham (20). [See also Kobayashi and Nomizu (13).] The first systematic study of completeness of a Riemannian manifold was made by Hopf and Rinow (7). [See also de Rham (20) and Kobayashi and Nomizu (13), which follows closely de Rham.] The statements of these facts are clear and we advise the reader to proceed assuming their validity.

We shall now explain exponential maps. At each point p of a complete Riemannian manifold M, we define a mapping of the tangent space $T_p(M)$ at p onto M in the following manner. If X is a tangent vector at p, we draw a geodesic $g(t)$ starting at p in the direction of X; we parametrize the geodesic in such a way that $g(0) = p$. If X has length a, then we map X into the point $g(a)$ of the geodesic. We denote this mapping by $\exp_p$ or simply $\exp$,

$$\exp_p \colon T_p(M) \to M$$

and call it the exponential map at p. Thus, $\exp_p$ maps a line in the tangent space $T_p(M)$ through its origin onto the geodesic of M through p in the direction of the line. Since every point q of M can be joined by a geodesic to p, $\exp_p$ maps $T_p(M)$ onto M.

To the reader who might wonder where the name "exponential map" originates, we point out that $\exp_p$ may be considered as a sort of generalization of the usual exponential function in the following way. In the theory of Lie groups, the exponential map of a Lie group G is a mapping of the tangent space $T_e(G)$ of G at the identity element e into G which sends a line through the origin of $T_e(G)$ onto the 1-parameter subgroup in the direction of the line. Since the tangent space $T_e(G)$ may be considered as the Lie algebra of G, the exponential map sends the Lie algebra of G into G. The ordinary exponential function is nothing but the exponential map of the multiplicative group of positive real numbers the Lie algebra of which is just the real line. If G is a compact Lie group, we can construct a Riemannian metric on G so that the 1-parameter subgroups are the geodesics through the identity element e and the exponential map as a Lie group is the exponential map at e as a Riemannian manifold. In passing, we might point out here that since every element of a compact G can be joined to e by a geodesic according to the aforementioned theorem of Hilbert, every element of a compact G lies on a 1-parameter subgroup of G.

We fix a point p of M. We shall now define the cut locus $C(p)$ of p. Take a geodesic $g(t)$, $0 \leq t < \infty$, starting at p. Then the first point on this geodesic where the geodesic ceases to minimize its arc length is called the *cut point of p along the geodesic $g(t)$*. To be more precise, we let A be the set of positive real numbers s such that the geodesic $g(t)$, $0 \leq t \leq s$, is minimizing—that is, $s = d(p, g(s))$. It is easy to see that either $A = (0, \infty)$ or $A = (0, r)$, where r is some positive number. If $A = (0, r)$, then $g(r)$ is the cut point of p along the geodesic $g(t)$. If $A = (0, \infty)$, then we say that p has no cut point along the geodesic $g(t)$. Thus, if q is a point on $g(t)$ which comes after the cut point $p' = g(r)$—that is, $q = g(s)$ with $s > r$—then we can find a geodesic from p to q which is shorter than $g(t)$. On the other hand, if q is a point which comes before the cut point p', then not only do we not find a shorter geodesic from p to q but we cannot have another geodesic from p to q of the same length. In other words, if q comes before p', then $g(t)$ is the *unique* minimizing geodesic joining p and q, for if there is another minimizing geodesic g' from p to q, then, by moving from p to q along g' and continuing from q to p' along g, we obtain a nongeodesic curve c from p to p' with an arc length equal to the distance $d(p, p')$. We choose a point p_1 on g' before q and also a point p_2 on g after q. Taking both p_1 and p_2 sufficiently close to q, we replace the portion of c from p_1 to p_2 by the minimizing geodesic joining p_1 and p_2 and obtain a curve from p to p' with an arc length less than the distance $d(p, p')$, which is absurd. The set of cut points of p is called the *cut locus* of p and is denoted by $C(p)$. Corresponding to the cut point p' of p along $g(t)$, we consider the point $X \in T_p(M)$, that is, the tangent vector X given by

$$g(t) = \exp_p(tX) \quad \text{and} \quad p' = \exp_p(X),$$

and call it the *cut point of p in $T_p(M)$* corresponding to the cut point p'.

If M is a compact Riemannian manifold, then its diameter is finite and no geodesic of length greater than the diameter is minimizing. If M is compact, on each geodesic starting from a point p we find the cut point of p. Correspondingly, in the tangent space $T_p(M)$ we find a cut point in every direction starting from p. It is quite reasonable to expect that the cut locus in $T_p(M)$ (not in M) is homeomorphic to a hypersphere in $T_p(M)$, provided that M is compact. To prove that this is actually the case, we have to show some kind of connectedness of the cut locus of p in $T_p(M)$, which we shall accomplish in the following theorem.

THEOREM 1.1: *Let S be the unit sphere in the tangent space $T_p(M)$ with center at the origin. We define a function f on S by assigning to each unit vector $X \in S$ the distance from p to the cut point in the direction of X. Then f is a continuous function.*

This theorem holds even for a noncompact manifold. However, if M is not compact, for some direction $X \in S$ the cut point may not exist. Then we set $f(X) = \infty$. Thus, we consider f as a mapping of S into $[0, \infty]$, where $[0, \infty]$ is the 1-point-compactification of $[0, \infty)$. To prove the theorem, we need properties of conjugate points as well as those of cut points; therefore, we postpone it to Section 4. We shall derive here some consequences from the theorem.

Assume that M is compact. As we have already stated, the cut locus of p in $T_p(M)$, denoted by $C^*(p)$, is homeomorphic to the unit sphere S in $T_p(M)$. Let E^* be the domain in $T_p(M)$ bounded by the cut locus $C^*(p)$. If we denote by E the image of E^* by $\exp_p$, then $\exp_p$ gives a homeomorphism of E^* onto E and we have, obviously, $M = E \cup C(p)$. Thus, M is a disjoint union of an open cell (that is, E) and a closed subset [that is, $C(p)$], which is a continuous image of an $(n-1)$-sphere [that is, $C^*(p)$], where $n = \dim M$. In particular, the cut locus $C(p)$ is connected.

The importance of cut loci lies in the fact that they inherit a number of topological properties of the manifold M. For example, let us consider the homotopy groups $\pi_k(M)$ and $\pi_k(C(p))$. We shall show that the natural mapping (that is, the imbedding) of $C(p)$ into M induces an isomorphism of $\pi_k(C(p))$ onto $\pi_k(M)$ for $k \leqq n-2$ and a homomorphism of $\pi_{n-1}(C(p))$ onto $\pi_{n-1}(M)$. Let S^k be the k-sphere and consider the element of $\pi_k(M)$ given by a continuous map $h:S^k \to M$. If $k \leqq n-1$, then the image $h(S^k)$ cannot cover all of M, and we may assume without loss of generality that $h(S^k)$ misses the point p. Since we can shrink $M - \{p\}$ into $C(p)$ by pushing it along the geodesic rays emanating from p toward $C(p)$, we can deform h to a continuous map $h':S^k \to C(p)$, showing that $\pi_k(C(p))$ is mapped onto $\pi_k(M)$ for $k \leqq n-1$. Let $f:S^k \to C(p)$ and assume that f can be extended to $g:B^{k+1} \to M$, where B^{k+1} is the solid ball with the boundary S^k. If $k+1 \leqq n-1$, then the image $g(B^{k+1})$ cannot cover M and we may assume that it misses the point p. We can deform q to $g':B^{k+1} \to C(p)$ as before, so that g' is an extension of f, showing that $\pi_k(C(p)) \to \pi_k(M)$ is an isomorphism for $k \leqq n-2$. We have a very natural element in $\pi_{n-1}(C(p))$, which belongs to the kernel $\pi_{n-1}(C(p)) \to \pi_{n-1}(M)$. Since the cut locus $C^*(p)$ in the tangent space $T_p(M)$ is homeomorphic with the $(n-1)$-sphere S^{n-1}, $\exp_p:C^*(p) \to C(p)$ gives an element of $\pi_{n-1}(C(p))$ that obviously belongs to the kernel.

Before we proceed, we should perhaps find out what cut loci look like for some simple Riemannian manifolds.

Examples

1. If M is an n-dimensional unit sphere, then the geodesics are the so-called great circles. If p is the north pole, then the cut locus $C(p)$ reduces to the south pole. The cut locus $C^*(p)$ in the tangent space $T_p(M)$ is the sphere of radius π with a center at the origin.

2. Identifying each point of the unit sphere S^n with its antipodal

point, we obtain an n-dimensional real projective space, which
we shall denote by M. The Riemannian metric of S^n induces a
Riemannian metric on M in a natural manner so that the projection of S^n onto M is a local isometry. If p is the point of M which
corresponds to the north and south poles of S^n, then the image
of the equator of S^n by the projection $S^n \to M$ is the cut locus
$C(p)$. In other words, $C(p)$ is the so-called hyperplane at the
infinity which is an $(n-1)$-dimensional real projective space.
The cut locus $C^*(p)$ in the tangent space $T_p(M)$ is the sphere of
radius $\pi/2$ with center at the origin and is a double covering
space of $C(p)$ under $\exp_p$.

3. In the euclidean plane R^2 with coordinate system (x, y),
consider the closed unit square $0 \leqq x, y \leqq 1$. By identifying $(x, 0)$
with $(x, 1)$ and also $(0, y)$ with $(1, y)$, we obtain a flat torus, which
we denote here by M. We are considering, of course, the Riemannian metric on M induced in a natural way from the euclidean
metric of R^2. Let p be the point with coordinates $(\frac{1}{2}, \frac{1}{2})$. Then the
cut locus $C^*(p)$ in $T_p(M)$ can be identified with the boundary of
the unit square under the natural identification of $T_p(M)$ with R^2.
The cut locus $C(p)$ consists of two intersecting circles which
generate the first homology group of the torus.

4. In the (x, y)-plane, consider a curve $y = f(x)$, $a \leqq x \leqq b$,
such that $f(a) = f(b) = 0$, $f(x) > 0$, for $a < x < b$, and such that
it is tangent to the lines $x = a$ and $x = b$. If we place this curve
in the (x, y, z)-space and revolve it about the x-axis, then we
obtain a smooth closed surface M which is topologically a sphere.
Let p be the point with coordinates $(a, 0, 0)$. Since p is invariant
under the revolution, which is a 1-parameter group of isometries,
so must be the cut locus $C(p)$. On the other hand, $M - C(p)$
must be connected. These two restrictions on $C(p)$ imply immediately that $C(p)$ reduces to a single point—that is, the point
with coordinates $(b, 0, 0)$. This fact applies in particular to the
ellipsoid $(x/a)^2 + (y/b)^2 + (z/c)^2 = 1$, with $b = c$. (On the general ellipsoid, the cut locus of any of the six poles is known to be
an arc on the longer of two principal ellipses through the given
pole and containing the opposite pole as midpoint. Also, on the

general ellipsoid, the cut locus of an umbilical point is the diametrically opposite umbilical point.)

The reader unfamiliar with complex or quaternionic projective spaces is advised to skip Example 5.

5. Let M be a complex (resp. quaternionic) projective space of real dimension $2n$ (resp. $4n$). If we normalize the metric so that the maximal sectional curvature is 1 (and hence the minimum sectional curvature is $\frac{1}{4}$), then the cut locus $C^*(p)$ in the tangent space $T_p(M)$ is the sphere of radius π with a center at the origin of $T_p(M)$. The cut locus $C(p)$ is the hyperplane at infinity—that is, the complex (resp. quaternionic) projective space of real dimension $2(n-1)$ (resp. $4(n-1)$) consisting of points at infinity with respect to p. The exponential map $\exp_p: C^*(p) \to C(p)$ defines a fiber bundle with fiber S^1 (resp. S^3) and is nothing but the so-called Hopf fibering.

The cut locus $C(p)$ is not a submanifold in general even if M is a homogeneous Riemannian manifold. (A Riemannian manifold is said to be homogeneous if the group of isometries acts transitively.) In Example 5, $C(p)$ is a nice submanifold because the group of isometries of M leaving p fixed is transitive on the directions at p—that is, transitive on the unit sphere in the tangent space $T_p(M)$ with a center at the origin. Even for a homogeneous Riemannian manifold it is usually difficult to determine the cut locus $C(p)$ of a point of p. It seems to be unknown if the cut locus $C(p)$ is, in general, a triangulable polyhedron, or something of that sort.

2. CONJUGATE LOCI

We shall begin with a geometric definition of conjugate point. As in Section 1, we denote by M a complete Riemannian manifold. Given a geodesic $g = g(t)$, $a \leqq t \leqq b$, consider a *variation* (or more precisely, a geodesic variation) of g—that is, a 1-parameter family of geodesics $g_s = g_s(t)$, $a \leqq t \leqq b$ and $-\epsilon < s < \epsilon$, such that $g = g_0$. For each fixed s, $g_s(t)$ describes a geodesic when t moves from a to b. Each variation gives rise to an *infinitesimal variation*, that is, a certain vector field defined along g. More explicitly, for each fixed t, consider the vector tangent to the curve $g_s(t)$, $-\epsilon < s < \epsilon$, at $s = 0$; it is a vector at the point $g(t)$ of g. Roughly speaking, the endpoints $g(a)$ and $g(b)$ of the geodesic g are said to be conjugate to each other if there is a variation g_s of g passing through these endpoints—that is, $g_s(a) = g(a)$ and $g_s(b) = g(b)$ for $-\epsilon < s < \epsilon$. But we do not actually require that g_s passes through these endpoints; it will be enough if the variation g_s passes these endpoints to infinitesimal order 1. We may now state the definition of conjugate points more precisely: The points $g(a)$ and $g(b)$ of g are said to be *conjugate* if there is a variation g_s which induces an infinitesimal variation vanishing at $t = a$ and $t = b$.

There is another equally geometric way of defining a conjugate point. Let p be a point of M and $\exp_p : T_p(M) \to M$ the exponential map defined in Section 1. A point X of $T_p(M)$—that is, a tangent vector at p—is called a conjugate point of p in $T_p(M)$ if $\exp_p$ is degenerate at X—that is, the Jacobian matrix of $\exp_p$ is singular at X. Its image, $p' = \exp_p (X)$ by $\exp_p$, is called a *conjugate point* of p along the geodesic $g(t) = \exp_p (tX)$.

It is easy to see that if p' is conjugate to p along g in the second sense, it is also so in the first sense. Since $\exp_p$ is degenerate at X, we can find a curve X_s through X in $T_p(M)$, such that $\exp_p$ annihilates the vector tangent to the curve X_s at $X = X_0$. We may also assume that for each s, the length of X_s is equal to that of X. Then $g_s(t) = \exp_p (tX_s/\|X\|)$ gives us a variation g_s of g with the desired property. The converse is a little harder to prove. Let p be the point $g(0)$ of a geodesic g and $p' = g(b)$ a conjugate point of p along g in the first sense. Let g_s be a variation of g such that the induced infinitesimal variation vanishes at p and p'. If we have exactly $q_s(0) = p$, then we have no difficulty in showing that p' is conjugate to p along g in the second sense. Considering the 1-parameter family of rays in $T_p(M)$ corresponding to the 1-parameter family of geodesics g_s through p, and taking the points with distance b from the origin on the rays, we obtain a curve X_s in $T_p(M)$ the tangent vector of which is annihilated by $\exp_p$ at X_0. This condition shows that X_0 is conjugate to p in $T_p(M)$, and, consequently, p' is conjugate to p in the second sense. It remains to prove that we can find a variation g_s such that not only the induced infinitesimal variation vanishes at p and p' but also $g_s(0) = p$. To achieve this proof, we give the third definition of conjugate point, which is less geometric but more analytic.

If Y is a vector field defined along geodesic g, then we denote by Y' its covariant derivative along g and by Y'' the second covariant derivative along g. For each geodesic g we have a naturally associated vector field along g—that is, the field of tangent vectors to g, which we shall denote by G. Since g is parametrized by its arc length, G is a unit vector field. By definition of a geodesic, we have $G' = 0$. A vector field Y along g is called a *Jacobi field* if it satisfies the following second-order, ordinary, linear, differential equation (called the Jacobi equation),

$$Y'' + R(Y, G)G = 0,$$

where $R(Y, G)$ denotes the curvature transformation determined by Y and G. (For any ordered pair of vectors at a point of M, the curvature determines a linear transformation of the tangent space at that point.) We say that points p and p' on a geodesic g are *conjugate* to each other if there is a nonzero Jacobi field Y along g which vanishes at p and p'. The equivalence of this definition with the first follows from the fact that a vector field Y defined along a geodesic g is a Jacobi field if and only if it is an infinitesimal variation of g.

We prove this fact as follows. Let g_s be a variation of g and set $f(s, t) = g_s(t)$. We denote by $D/\partial t$ and $D/\partial s$ the covariant differentiations with respect to t and s, respectively. Since g_s is a geodesic for each s, we have

$$\frac{D}{\partial t} \frac{\partial f}{\partial t} = 0.$$

Hence,

$$0 = \frac{D}{\partial s} \frac{D}{\partial t} \frac{\partial f}{\partial t} = \frac{D}{\partial t} \frac{D}{\partial s} \frac{\partial f}{\partial t} + R\left(\frac{\partial f}{\partial t}, \frac{\partial f}{\partial s}\right) \frac{\partial f}{\partial t}$$

$$= \frac{D}{\partial t} \frac{D}{\partial t} \frac{\partial f}{\partial s} + R\left(\frac{\partial f}{\partial t}, \frac{\partial f}{\partial s}\right) \frac{\partial f}{\partial t},$$

showing that an infinitesimal variation is a Jacobi field. (The conscientious reader will find the preceding calculation somewhat ambiguous; since f is not a one-to-one mapping into M, neither $\partial f/\partial t$ nor $\partial f/\partial s$ is well defined as a vector field on the image of f. Returning to the definition of covariant differentiation and curvature, we can explain the ambiguities. (The use of differential forms and the so-called structure equations of Cartan is a more satisfactory proof, but the use of such machineries is inappropriate in this article and we have to be satisfied with the preceding proof.)

Conversely, let Y be a Jacobi field along g. We choose two parameter values a and b close to each other so that $g(a)$ and $g(b)$ are in a convex neighborhood (any two points of which can be joined by a unique minimizing geodesic). We choose a curve $g_s(a)$, $-\epsilon < s < \epsilon$, through the point $g(a)$ which is tangent to Y at $g(a)$. Similarly, we let $g_s(b)$, $-\epsilon < s < \epsilon$, be a curve through $g(b)$ tangent to Y at $g(b)$. We can also arrange it so that the distance $d(g_s(a), g_s(b))$ is equal to $d(g(a), g(b))$ for each s. Letting $g_s = g_s(t)$ be the unique minimizing geodesic joining $g_s(a)$ and $g_s(b)$, we find that the infinitesimal variation induced by the variation g_s of g coincides with Y. The construction shows that if Y vanishes at $g(a)$, we can find a variation g_s with the additional property $g_s(a) = g(a)$. We have thus shown simultaneously the equivalence of the first and second definitions.

Perhaps the most important method of obtaining Jacobi fields is by infinitesimal isometries. Let φ_s be a 1-parameter group of isometries of M. Given a geodesic g, we obtain a variation g_s by $g_s = \varphi_s(g)$. The induced infinitesimal variation is nothing but the infinitesimal transformation (restricted to g) which generates φ_s. If p and p' are points on g which are left fixed by φ_s, then p and p' are conjugate to each other, provided that the Jacobi field along g induced by φ_s is nontrivial.

Let M^* be a Riemannian covering manifold of M; it is a covering space of M with the naturally induced Riemannian structure. Let g^* be a geodesic in M^* and g its image on M under the natural projection of M^* onto M. Then two points on g^* are conjugate along g^* if and only if their images on g are conjugate along g. This relation is evident when we consider Jacobi equations.

We shall now consider a few simple examples.

Examples

1. If M is an n-dimensional unit sphere and p is the north pole, then both p itself and the south pole p' are conjugate to p along any geodesic—that is, great circle—through p. We see easily that no other points are conjugate to p.

2. Let M be the real projective space of dimension n as in Example 2, Section 1, and let p be a point of M. Then p itself is conjugate to p along any geodesic through p and no other points are conjugate to p, because a sphere is a Riemannian covering manifold of M.

3. Let M be a flat torus as in Example 3, Section 1. Since the universal covering manifold of M is a euclidean space and there are no conjugate points in a euclidean space, there are no conjugate points on M.

4. Let M be a surface of revolution as in Example 4, Section 1. Since the revolution about the x-axis, which is a 1-parameter group of isometries of M, induces an infinitesimal variation vanishing at $p = (a, 0, 0)$ and at $q = (b, 0, 0)$ (and at no other points), p and q are conjugate to each other. Because the cut locus $C(p)$ of p consists of q only, and because along any geodesic starting at p the first conjugate point of p generally never comes before the cut point of p (as we shall see in Section 4), the locus of first conjugate points of p consists of q only. This condition applies in particular to the ellipsoid $(x/a)^2 + (y/b)^2 + (z/c)^2 = 1$ with $b = c$. [On the general ellipsoid, the situation is more complicated. See the picture in Struik (22), p. 143, and see also Braunmühl (3).]

5. If M is a complex or quaternionic projective space as in Example 5 of Section 1, the first conjugate locus of a point p coincides with the cut locus $C(p)$ of p.

In spite of some of the preceding examples, the cases where the cut locus $C(p)$ coincides with the first conjugate locus of p are exceptional rather than general. See Section 5 for further discussion.

3. CURVATURE AND CONJUGATE POINTS

By looking at the Jacobi equation, we can see it is natural to expect a close relationship between the curvature and the distribution of conjugate points. The Jacobi equation is a tensorial analogue of a differential equation of the type

$$y'' + K(x)y = 0,$$

which has been studied by Sturm. For another differential equation

$$y'' + L(x)y = 0$$

of the same type such that

$$K(x) \leqq L(x),$$

the comparison theorem of Sturm says that if $u(x)$ is a solution of the first equation having m zeros in the interval $a \leqq x \leqq b$, and if $v(x)$ is a solution of the second equation with

$$u(a) = v(a) \quad \text{and} \quad u'(a) = v'(a),$$

then $v(x)$ has at least m zeros in the same interval. If $K(x)$ is a positive constant function K, then a solution of the first equation is of the form $u(x) = C_1 \sin \sqrt{K}x + C_2 \cos \sqrt{K}x$ and, hence, the distance between any two consecutive zeros of a solution $v(x)$ of the second equation is, at most, $\pi/\sqrt{K}$. On the other hand, if $L(x)$ is a positive constant function L, then the distance between two consecutive zeros of a solution $u(x)$ of the first equation is at least $\pi/\sqrt{L}$. Analogously, we have the following theorem.

THEOREM 3.1 (BONNET): *Let g be a geodesic in M and $K(P)$ the sectional curvature of a plane section P tangent to g. If*

$$0 < L \leqq K(P) \leqq H$$

for all such plane sections P, then the distance d along g of any two consecutive conjugate points satisfies the following inequalities:

$$\frac{\pi}{\sqrt{H}} \leqq d \leqq \frac{\pi}{\sqrt{L}}.$$

The theorem follows from the comparison theorem of Sturm. The reader will find the proof for surfaces in a number of standard textbooks on elementary differential geometry of surfaces. The proof for higher dimensional manifolds may be reduced to that for surfaces by Synge's lemma which states: If V is a (2-dimensional) surface imbedded in M and g is a geodesic of M lying in V, then the Gaussian curvature of V at a point of g is less than, or equal to, the sectional curvature of M for the tangent plane to V at that point; the equality holds when and only when the field of tangent planes to V is parallel along g in M. This lemma of Synge is an immediate consequence of the equations of Gauss and Coddazi for V imbedded in M, but we shall prove the comparison theorem of Rauch which implies immediately Bonnet's theorem.

THEOREM 3.2 (RAUCH'S COMPARISON THEOREM): *Let M and N be Riemannian manifolds of the same dimension n. Let $g = g(t)$ be a geodesic of M and X a Jacobi field along g. Let $h = h(t)$ be a geodesic of N and Y a Jacobi field along h. Assume that: (1) X is perpendicular to g and vanishes at $g(0)$, Y is perpendicular to h and vanishes at $h(0)$; (2) X' and Y' have the same length at $t = 0$; (3) neither g nor h has conjugate points in the interval $(0, b)$; (4) for each t in the interval $(0, b)$,*

$$K_M(P) \geqq K_N(Q),$$

where $K_M(P)$ (resp. $K_N(Q)$) is the sectional curvature by an arbitrary tangent plane P (resp. Q) at $g(t)$ tangent to g (resp. at $h(t)$ tangent to h). Under these four assumptions, Y at $h(b)$ is longer than, or equal to, X at $g(b)$.

Before we proceed with the proof, we remark that Theorem 3.1 follows from Theorem 3.2 if we take M to be the sphere of radius H and N to be M of Theorem 3.1, and if we then take N to be the sphere of radius L.

Proof: Let $u(t)$ and $v(t)$ be the square of the lengths of X at $g(t)$ and Y at $h(t)$, respectively. We wish to show $u(b) \leqq v(b)$. It is sufficient to prove

(3.1) $$\lim_{t \to 0} \frac{u(t)}{v(t)} = 1;$$

(3.2) $$\left(\frac{u(t)}{v(t)}\right)' \leq 0,$$

that is, $u(t)/v(t)$ is decreasing. Since $u' = 2(X, X')$, $v' = (Y, Y')$, $u'' = 2(X', X') + 2(X, X'')$, and $v'' = 2(Y', Y') + 2(Y, Y''')$, we apply l'Hospital's rule twice to $u(t)/v(t)$ and obtain (3.1).

Proving (3.2) is equivalent to proving $u'(t)/u(t) \leq v'(t)/v(t)$. We fix a parameter value s, $0 < s < b$. By the third assumption, $u(s) \neq 0$ and $v(s) \neq 0$. We may thus set $U = X/u(s)^{1/2}$ and $V = Y/v(s)^{1/2}$. Since X and Y are Jacobi fields, so are U and V. We have, therefore,

$$\frac{u'(s)}{u(s)} = (U, U)'(s) = \int_0^s (U, U)'' \, dt$$

$$= 2 \int_0^s [(U', U') - K_M(P)(U, U)] \, dt,$$

where $K_M(P)$ denotes the sectional curvature of the planes P spanned by U and the vectors tangent to the geodesic g. Similarly,

$$\frac{v'(s)}{v(s)} = 2 \int_0^s [(V', V') - K_N(Q)(V, V)] \, dt.$$

Let $\bar{U}$ be a vector field defined along g such that

$$(\bar{U}, \bar{U})(t) = (V, V)(t), \qquad (\bar{U}', \bar{U}')(t) = (V', V')(t).$$

(We may establish the existence of such a $\bar{U}$ by choosing an isomorphism of the normal space to g at $g(0)$ onto the normal space to h at $h(0)$, extending it to an isomorphism of the normal space to g at $g(t)$ onto the normal space to h at $h(t)$, and then considering the vector field along g corresponding to V.) We shall show that

$$\frac{u'(s)}{u(s)} \leq 2 \int_0^s [(\bar{U}', \bar{U}') - K_M(\bar{P})(\bar{U}, \bar{U})] \, dt \leq \frac{v'(s)}{v(s)}.$$

The second inequality follows from the fourth assumption and from the preceding integral expression for $v'(s)/v(s)$. To prove the first inequality, we need the following lemma.

LEMMA 1: *If X and Y are Jacobi fields along a geodesic g, then*

$$(X, Y') - (X', Y) = const.$$

If, moreover, X and Y vanish at a point simultaneously, then

$$(X, Y') - (X', Y) = 0.$$

A simple direct calculation shows that the derivative of $(X, Y') - (X', Y)$ vanishes identically. The second statement of the lemma follows trivially from the first.

Using Lemma 1, we shall prove the following lemma which will obviously imply the first inequality.

LEMMA 2: *Let* $g = g(t)$, $0 \leqq t \leqq s$, *be a geodesic without conjugate point of* $g(0)$ *in the interval* $0 \leqq t \leqq s$. *Let* U *be a Jacobi field along* g *perpendicular to* g, *and* $\overline{U}$ *a vector field along* g *perpendicular to* g. *If* $U(0) = \overline{U}(0) = 0$, *and if* $U(s) = \overline{U}(s)$, *then*

$$\int_0^s \left[(U', U') - K_M(P)(U, U) \right] dt$$

$$\leqq \int_0^s \left[(\overline{U}', \overline{U}') - K_M(\overline{P})(\overline{U}, \overline{U}) \right] dt,$$

where $K_M(P)$ *and* $K_M(\overline{P})$ *denote the sectional curvatures of the planes* P *spanned by* U *and the vectors tangent to* g, *and of the planes* $\overline{P}$ *spanned by* $\overline{U}$ *and the vectors tangent to* g. *The equality holds only when* $U = \overline{U}$.

Consider the vector space of Jacobi fields Z perpendicular to g and vanishing at $g(0)$. Being a solution of a second order differential equation, Z is uniquely determined by its initial conditions $Z(0)$ and $Z'(0)$. Since $Z(0) = 0$ and $Z'(0)$ must be perpendicular to g, the vector space of these Z's is of dimension $n - 1$. Let $Z_1, \cdots, Z_{n-1}$ be a basis for this vector space. Because there exists no conjugate point of $g(0)$ in the interval $(0, s)$, $Z_1, \cdots, Z_{n-1}$ are linearly independent at each point $g(t)$, $0 < t \leqq s$. We may write, therefore,

$$U = a_1 Z_1 + \cdots + a_{n-1} Z_{n-1},$$

$$\overline{U} = f_1 Z_1 + \cdots + f_{n-1} Z_{n-1},$$

where $a_1, \cdots, a_{n-1}$ are constants and $f_1, \cdots, f_{n-1}$ are functions. We have

$$(\overline{U}', \overline{U}') = (\Sigma f_i' Z_i, \Sigma f_i' Z_i) + 2(\Sigma f_i' Z_i, \Sigma f_i Z_i')$$
$$+ (\Sigma f_i Z_i', \Sigma f_i Z_i'),$$
$$-(R(\overline{U}, G)G, \overline{U}) = -\Sigma f_i(R(Z_i, G)G, \overline{U}) = \Sigma f_i(Z_i'', \overline{U})$$
$$= (\Sigma f_i Z_i'', \Sigma f_i Z_i),$$

where G denotes the vector field tangent to g, and we also have

$$(\Sigma f_i Z_i, \Sigma f_i Z_i')' = (\Sigma f_i Z_i, \Sigma f_i Z_i') + (\Sigma f_i Z_i', \Sigma f_i Z_i')$$
$$+ (\Sigma f_i Z_i, \Sigma f_i' Z_i') + (\Sigma f_i Z_i, \Sigma f_i Z_i'').$$

Combining these three equalities and using Lemma 1, we obtain

$$\int_0^s [(\overline{U}', \overline{U}') - K_M(\overline{P})(\overline{U}, \overline{U})]\, dt = \int_0^s (\Sigma f_i' Z_i, \Sigma f_i' Z_i)\, dt$$
$$+ (\Sigma f_i Z_i, \Sigma f_i Z_i')_s.$$

In the preceding calculation, we replace $\overline{U}$ by U, $\overline{P}$ by P, and f_i by a_i, and observe that $a_i' = 0$. Then we obtain

$$\int_0^s [(U', U') - K_M(P)(U, U)]\, dt = \int_0^s (\Sigma a_i Z_i, \Sigma a_i Z_i')_s.$$

Lemma 2 follows from the two preceding equalities and from the observation $a_i = f_i(s)$.

4. RELATIONS BETWEEN CUT POINTS AND CONJUGATE POINTS

The first general statement we can make about relations between cut points and conjugate points is that along any geodesic g starting at a point, say p, the cut point of p comes no later than the first conjugate point of p. In other words, we prove the following theorem.

THEOREM 4.1: *A geodesic g starting at a point p does not minimize distance to p beyond the first conjugate point of p.*

Proof: Let $g(c)$ be the first conjugate point of p and let $c < b$. We wish to show that the distance $d(g(0), g(b))$ is less than the arc length b of g from $t = 0$ to $t = b$, which is intuitively obvious. Suppose we have a variation $g_s = g_s(t)$ of the geodesic g such that $g_s(0) = g(0)$ and $g_s(c) = g(c)$. We shall show that there is a shorter way to go from $p = g(0)$ to $g(b)$ than along g. We fix a small e and first consider a broken path from $p = g_e(0)$ to $g_e(c) = g(c)$ via g_e, and then from $g(c)$ to $g(b)$ via g, which has length $c + (b - c) = b$. Smoothing out the corner of this path at $g(c)$, we obtain a shorter path from p to $g(b)$. However, in general we have only an infinitesimal variation or a Jacobi field vanishing at $g(0)$ and $g(c)$ but not necessarily a variation with the fixed points $g(0)$ and $g(c)$. Therefore, the rigorous proof is more delicate. We first prove the following lemma.

LEMMA 1: *Let g, b, and c be defined as previously. Then there exists a vector field X along g such that:* (1) X *is perpendicular to g;* (2) X *vanishes at $t = 0$ and at $t = b$;* (3) $\int_0^b [(X', X') - (R(X, G)G, X)] \, dt < 0$, *where G denotes the vector field tangent to g.*

Proof: Let Y be a nonzero Jacobi field along g vanishing at $t = 0$ and $t = c$. Take a small positive number d such that $g(c - d)$ has no conjugate points along g between $g(c - d)$ and $g(c + d)$ and such that there is a Jacobi field Z along g with $Z(c - d) = Y(c - d)$ and $Z(c + d) = 0$. (It is sufficient to take a positive number d such that $g(c - d)$ and $g(c + d)$ are in a convex neighborhood U of $g(c)$ such that each point of U has a normal coordinate neighborhood containing U.) We define a vector field X along $g = g(t)$, $0 \leq t \leq b$, as follows:

$$X = Y \quad \text{for } 0 \leq t \leq c - d,$$
$$X = Z \quad \text{for } c - d \leq t \leq c + d,$$
$$X = 0 \quad \text{for } c + d \leq t \leq b.$$

To simplify the calculation, we introduce the following notation: If W is a vector field along f, we set

$$I_r^s(W) = \int_r^s [(W', W') - (R(W, G)G, W)] \, dt.$$

Integrating

$$0 = (Y'' + R(Y, G)G, Y) = (Y'', Y) + (R(Y, G)G, Y)$$
$$= (Y', Y)' - (Y', Y') + (R(Y, G)G, Y)$$

from $t = 0$ to $t = c$, we obtain

$$I_0^c(Y) = 0.$$

Thus,

$$I_0^b(X) = I_0^b(X) - I_0^c(Y) = I_0^{c-d}(Y) + I_{c-d}^{c+d}(Z)$$
$$-I_0^{c-d}(Y) - I_{c-d}^c(Y)$$
$$= I_{c-d}^{c+d}(Z) - I_{c-d}^c(Y).$$

Let $\overline{Y}$ be the vector field along g from $t = c - d$ to $t = c + d$ defined by

$$\overline{Y} = Y \quad \text{for } c - d \leqq t \leqq c,$$
$$\overline{Y} = 0 \quad \text{for } c \leqq t \leqq c + d.$$

Applying Lemma 2 of the preceding section to $\overline{Y}$ and Z, we have

$$I_{c-d}^{c+d}(Z) < I_{c-d}^{c+d}(\overline{Y}) = I_{c-d}^c(Y).$$

Hence,

$$I_0^b(X) < 0.$$

Finally, we have to show that X is perpendicular to g—that is, Y and Z are perpendicular to g. It is sufficient to prove that, in general, a Jacobi field V along g which is perpendicular to g at two points is perpendicular to g everywhere. Consider the function (V, G) of t, which vanishes for two distinct values of t. Differentiating twice, we have

$$(V, G)'' = (V'', G) = (R(V, G)G, G) = 0.$$

Hence, (V, G) is identically zero, which completes the proof of Lemma 1.

In the intuitive but incomplete proof of Theorem 4.1 given previously, we constructed a path from $p = g(0)$ to $g(b)$ which is shorter than the geodesic segment $g(t)$, $0 \leqq t \leqq b$. The vector field X in Lemma 1 may be considered an infinitesimal analogue of that path.

Let X be the vector field along g given in Lemma 1. Let $g_s = g_s(t)$ be a 1-parameter family of curves from $p = g(0)$ to $g(b)$, such that (1) $g_0 = g$, and (2) g_s induces X—that is, for each fixed t, $X(t)$ is the vector tangent to the curve described by $g_s(t)$, $-\epsilon < s < \epsilon$. We remark that for $s \neq 0$ g_s need not be a geodesic. For each fixed s, let $L(s)$ be the length of the curve $g_s = g_s(t)$, $0 \leq t \leq b$. (For $s \neq 0$, g_s is not necessarily parametrized by arc length and $L(s)$ may be different from b.) For completion of the proof of Theorem 4.1, it is sufficient to show that the function L of s attains a local maximum at $s = 0$.

LEMMA 2: $L'(0) = 0$ and $L''(0) = I_0^b(X) < 0$.

Proof: The proof may be given by calculation in a manner similar to that in Section 2. (Here, again, a more satisfactory proof can be given with the use of differential forms and Cartan's structure equations.) As previously, we set $f(s, t) = g_s(t)$ and denote by $D/\partial t$ and $D/\partial s$ the covariant differentiations with regard to t and s respectively. Then

$$L(s) = \int_0^b \left(\frac{\partial f(s, t)}{\partial t}, \frac{\partial f(s, t)}{\partial t} \right)^{1/2} dt.$$

To differentiate $L(s)$ with respect to s at $s = 0$, we first differentiate the integrand with respect to s. We have

$$\frac{\partial}{\partial s} \left(\frac{\partial f}{\partial t}, \frac{\partial f}{\partial t} \right)^{1/2} = \left(\frac{D}{\partial s} \frac{\partial f}{\partial t}, \frac{\partial f}{\partial t} \right) \left(\frac{\partial f}{\partial t}, \frac{\partial f}{\partial t} \right)^{-1/2}.$$

At $s = 0$, $[(\partial f/\partial t), (\partial f/\partial t)] = 1$, because $g = g_0$ is a geodesic. We can also substitute $(D/\partial t)(\partial f/\partial s)$ for $(D/\partial s)(\partial f/\partial t)$. Hence, integrating by parts, we obtain

$$L'(0) = \int_0^b \left(\frac{D}{\partial t} \frac{\partial f}{\partial s}, \frac{\partial f}{\partial t} \right)_{s=0} dt$$

$$= \left(\frac{\partial f}{\partial s}, \frac{\partial f}{\partial t} \right)_{s=0} \Big|_{t=0}^{t=b} - \int_0^b \left(\frac{\partial f}{\partial s}, \frac{D}{\partial t} \frac{\partial f}{\partial t} \right)_{s=0} dt$$

$$= (X(b), G(b)) - (X(0), G(0)) - \int_0^b (X, G') dt.$$

Since $X(b) = 0$ and $X(0) = 0$, and because g is a geodesic—that is, $G' = 0$—we have $L'(0) = 0$. Referring to the preceding calculation and also to the calculation in Section 2, the reader should easily be able to obtain the equality $L''(0) = I_0^b(X)$. By Lemma 1, $I_0^b(X) < 0$, and the proof of Theorem 4.1 is complete.

Another theorem on cut points and conjugate points follows.

THEOREM 4.2: *If $g(b)$ is the cut point of a point $p = g(0)$ along a geodesic $g = g(t)$, $0 \leqq t < \infty$, then one of the following (or possibly both) holds: (1) $g(b)$ is the first conjugate point of $g(0)$ along g; (2) there exist at least two minimizing geodesics from $g(0)$ to $g(b)$.*

Proof: Let a_1, a_2, $\cdots$ be a monotone decreasing sequence of real numbers converging to b. For each k, let $\exp tX_k$, $0 \leqq t \leqq b_k$, be a minimizing geodesic from $g(0)$ to $g(a_k)$, where X_k is a unit tangent vector at $g(0)$ and b_k is the distance $d(g(0), g(a_k))$. Write $g(t) = \exp tX$, where X is the unit vector tangent to g at $g(0)$. Since $g(b)$ is the cut point of $g(0)$ along g and $a_k > b$, we have

$$X \neq X_k \quad \text{and} \quad a_k > b_k.$$

Since $b_k = d(g(0), g(a_k))$, we have

$$b = \lim b_k.$$

Hence, the set of vectors $b_k X_k$ is contained in a compact subset of the tangent space $T_{g(0)}(M)$. By taking a subsequence if necessary, we may assume that $b_1 X_1$, $b_2 X_2$, $\cdots$ converges to a vector of length b—say bY—where Y is a unit vector. Then $\exp tY$, $0 \leqq t \leqq b$, is a minimizing geodesic from $g(0)$ to $g(b)$, because

$$\exp bY = \lim \exp b_k X_k = \lim g(a_k) = g(b).$$

If $X \neq Y$, then we have two minimizing geodesics, $\exp tX$ and $\exp tY$ from $g(0)$ to $g(b)$. If $X = Y$, then $g(b)$ is conjugate to $g(0)$ along g. In fact, assume the contrary. Then $\exp: T_{g(0)}(M) \rightarrow M$ is a diffeomorphism of a neighborhood U of bX in $T_{g(0)}(M)$ onto a neighborhood of $g(b)$ in M. If k is large enough so that both $a_k X$ and $b_k X_k$ are in U, then $\exp a_k X = g(a_k) = \exp b_k X_k$, and, hence, $a_k X = b_k X_k$, which is a contradiction. By Theorem 4.1, $g(b)$ must be the *first* conjugate point of $g(0)$ along g, which completes the proof.

From Theorem 4.2 it follows rather easily that if $g(b)$ is the cut point of $g(0)$ along g, then $g(0)$ is the cut point of $g(0)$ along g (in the reverse direction).

We are now in position to prove Theorem 1.1 as we promised in Section 1. With the notations used in Theorem 1.1, we wish to prove that the function f defined on the sphere S is continuous. Actually we shall prove a slightly stronger statement. At each point p of M we consider the unit sphere S_p in the tangent space $T_p(M)$ and let $S(M)$ be the unit sphere bundle—that is, $S(M) = \bigcup_p S_p$.

THEOREM 4.3: *Let $f: S(M) \to [0, \infty]$ be the function defined as follows: For each unit vector $X \in S(M)$ at p, $f(X)$ is the distance from p to its cut point along the geodesic issued in the direction of X. Then f is continuous.*

Proof: Theorem 1.1 asserts that the restriction of f to each S_p is continuous. To prove the theorem, we assume that f is not continuous at $X \in S_p$ and let $X_1, X_2, \cdots$ be a sequence of points of $S(M)$ converging to X, such that $f(X) \neq \lim f(X_k)$. By taking a subsequence if necessary, we may assume that $\lim f(X_k)$ exists in $(0, \infty)$.

We first consider the case $f(X) > \lim f(X_k)$. We set

$$a_k = f(X_k), \qquad a = \lim a_k.$$

Let $T(M)$ denote the tangent bundle of M, that is, $T(M) = \bigcup_p T_p(M)$. We define a mapping $E: T(M) \to M \times M$ by

$$E(Z) = (\pi(Z), \exp Z),$$

where π is the projection from $T(M)$ onto M, $\pi(T_p(M)) = p$. Since $f(X) > a$, $\exp aX$ cannot be conjugate to p along the geodesic $\exp tX$. Hence, E maps a neighborhood, say U, of aX diffeomorphically onto a neighborhood of $(\pi(X), \exp aX)$. We may assume, by omitting a finite number of $a_k X_k$ if necessary, that all $a_k X_k$ are in U. Then $\exp a_k X_k$ cannot be conjugate to $\pi(X_k)$ along the geodesic $\exp tX_k$. By Theorem 4.2, there is another minimizing geodesic $\exp tY_k$ from $\pi(X_k)$ to $\exp a_k X_k$, where Y_k is a unit vector at $\cdot \pi(X_k)$ such that $Y_k \neq X_k$ and $\exp a_k Y_k = \exp a_k X_k$. Since $E: U \to M \times M$ is injective, $a_k Y_k$ is not in U. By taking a subsequence if necessary, we may assume that $Y_1, Y_2, \cdots$ converges to a unit vector, say Y, at p. Then aY is the limit of $a_1 Y_1, a_2 Y_2, \cdots$ and does not lie in U. We have

$$\exp aY = \exp (\lim a_k Y_k) = \lim (\exp a_k Y_k) = \lim (\exp a_k X_k)$$
$$= \exp (\lim a_k X_k) = \exp aX.$$

Hence, both $\exp tX$ and $\exp tY$ are minimizing geodesics from p to $\exp aX = \exp aY$. Consequently, if b is any number greater than a, the geodesic $\exp tX$, $0 \leq t \leq b$, is not minimizing, in contradiction to the assumption $f(X) > a$.

Next we consider the case $f(X) < \lim f(X_k)$. As before, we set $a_k = f(X_k)$ and $a = \lim a_k$. Then

$$d(p, \exp aX) = \lim d(\pi(X_k), \exp a_k X_k) = \lim a_k = a,$$

showing that $\exp tX$, $0 \leq t \leq a$, is a minimizing geodesic in contradiction to the assumption $f(X) < a$. The proof of Theorem 4.3 is thus complete.

As an immediate consequence, we have the following corollary.

COROLLARY: *The distance $d(p, C(p))$ between a point p and its cut locus $C(p)$ is a continuous function of p.*

For a special point of $C(p)$, we can sharpen the result of Theorem 4.2 as follows.

THEOREM 4.4: *Let q be a point on the cut locus $C(p)$ of p, which is closest to p. Then either q is conjugate to p with respect to a minimizing geodesic joining p and q, or q is the midpoint of a geodesic starting and ending at p.*

Proof: Assuming that q is not conjugate to p, we let g and g' be two minimizing geodesics from p to q and b be the distance $d(p, q)$. We let K be a cone formed by geodesics of length b issuing from p and neighboring g. Similarly, for g' we consider a cone K'. The endpoints of the family of geodesics defining K give a hypersurface through q, with a tangent space at q perpendicular to g. The endpoints of the geodesics defining K' give another hypersurface through q, with a tangent space at q perpendicular to g'. We assume that g and g' meet at q with an angle not equal to π. Then the two tangent hyperplanes at q intersect (they do not coincide), as do the two hypersurfaces. It follows that K and K' intersect. We let r be a point in $K \cap K'$ near q. Then r is joined by two geodesics, one neighboring g and the other neighboring g'

and each being shorter than g and g'. Thus, r is a cut point of p closer to p than is the point q, in contradiction to the choice of q.

Taking a special point p, we shall go one step further.

THEOREM 4.5: *Let p be a point such that $d(p, C(p))$ is the smallest, and let q be a point on $C(p)$ which is closest to p. Then either q is conjugate to p with respect to a minimizing geodesic joining p and q, or q is the midpoint of a closed geodesic starting and ending smoothly at p.*

Proof: We assume that q is not conjugate to p. Applying Theorem 4.4, we see that q is the midpoint of a geodesic starting and ending at p. To see that the geodesic returns to p with angle 0, we reverse the roles of p and q in the proof of Theorem 4.4.

5. CONCLUDING REMARKS

Let M be a compact, simply connected Riemannian manifold the sectional curvature of which, $K(P)$, satisfies the inequalities $0 < K(P) \leq 1$. Let p be a point of M such that the distance from p to its cut locus $C(p)$ is the smallest. Let q be a point on $C(p)$ which is closest to p. According to Theorem 4.5, two possibilities exist. If q is conjugate to p, then $d(p, C(p)) \geq \pi$ by Theorem 3.1. If p and q are opposite points of a closed geodesic, then $d(p, C(p)) \geq \pi$ under the additional assumption that the dimension of M is even [see Klingenberg (8)]. If we assume $\frac{1}{4} \leq K(P) \leq 1$, then $d(p, C(p)) \geq \pi$, even if the dimension of M is odd [see Klingenberg (10 and 11)].

The inequality $d(p, C(p)) \geq \pi$ is essential in proving the so-called sphere theorem, which states that if M is a compact, simply connected Riemannian manifold with $\frac{1}{4} < K(P) \leq 1$, then M is homeomorphic with a sphere. The proof may be divided into the following three lemmas.

LEMMA 1: *If $\frac{1}{4} < K(P) \leq 1$, then $d(p, C(p)) \geq \pi$ for every $p \in M$.*

LEMMA 2: *Let h and a be positive numbers such that*

$$\tfrac{1}{4} < h \leq K(P) \leq 1 \quad and \quad \frac{\pi}{2\sqrt{h}} < a < \pi.$$

If p and q are points of M such that the distance $d(p, q)$ is equal to the diameter of M, then $d(p, x) < a$ or $d(q, x) < a$ for every $x \in M$.

LEMMA 3: *If there exist two points p and q and a positive number a such that*

(5.1) $d(p, C(p)) > a \quad and \quad d(q, C(q)) > a;$

(5.2) $d(p, x) < a \quad or \quad d(q, x) < a \quad for \ every \ x \in M,$

then M is homeomorphic with a sphere.

It is evident that Lemmas 1 through 3 imply the sphere theorem. For the proofs of Lemmas 2 and 3, we refer the reader to Tsukamoto (27) and Klingenberg (8), respectively. References for Lemma 1 have been given previously. The second lemma was proved originally by Berger (1) using results of Toponogov (24 and 25). For the proof of the sphere theorem, see Toponogov (26). By analyzing the distribution of conjugate points and by using Morse theory instead of the method we have described, it is possible to prove that, under the same assumption as in the sphere theorem, M is a homotopy sphere [see Klingenberg (12)]. For an improvement of Rauch's comparison theorem and further properties of cut loci, see Berger (2).

We have seen some examples in which the cut locus $C(p)$ of a point p coincides with the first conjugate locus of p. This phenomenon occurs for every simply connected Riemannian symmetric space [see Crittenden (5)].

BIBLIOGRAPHY

1. Berger, M., "Les variétés Riemanniennes ($\frac{1}{4}$)-pincées," *Annali della Scuola Normale Sup. di Pisa, Ser. III* 14 (1960), pp. 161–70.

2. Berger, M., "An extension of Rauch's metric comparison theorem and some applications," *Illinois Journal of Mathematics*, 6 (1962), pp. 700–12.

3. Braunmühl, A. v., "Über Enveloppen geodätischer Linien," *Math. Ann.*, 14 (1879), pp. 557–66.

4. Cartan, É., *Leçons sur la géométrie des espaces de Riemann*. Paris: Gauthier-Villars, 1928, 1946.

5. Crittenden, R., "Minimum and conjugate points in symmetric spaces," *Canadian Journal of Mathematics*, 14 (1962), pp. 320–28.

6. Hilbert, D., "Über das Dirichletsche Prinzip," *J. Reine Angew. Math.*, 129 (1905), pp. 63–67.

7. Hopf, H., and W. Rinow, "Über den Begriff der vollständigen differential geometrischen Fläche," *Comm. Math. Helv.*, 3 (1931), 209–25

8. Klingenberg, W., "Contributions to Riemannian geometry in the large," *Ann. of Mathematics*, 69 (1959), 654–66.

9. Klingenberg, W., "Über kompakte Riemannsche Mannigfaltigkeiten," *Math. Ann.*, 137 (1959), 351–61.

10. Klingenberg, W., "Über Riemannsche Mannigfaltigkeiten mit positiver Krümmung," *Comm. Math. Helv.*, 35 (1961), pp. 47–54.

11. Klingenberg, W., "Über Riemannsche Mannigfaltigkeiten mit nach oben beschränkter Krümmung," *Annali di Mat.*, 60 (1963), pp. 49–60.

12. Klingenberg, W., "Manifolds with restricted conjugate locus," *Ann. of Math.*, 78 (1963), pp. 527–47.

13. Kobayashi, S., and K. Nomizu, *Foundations of Differential Geometry*. New York: Wiley-Interscience, 1963.

14. Morse, M., *The Calculus of Variations in the Large*. Providence, R.I.: American Mathematical Society, 1934. (Reprinted 1965.)

15. Myers, S. B., "Connections between differential geometry and topology I," *Duke Mathematics Journal*, 1 (1935), 376–91.

16. Myers, S. B., "Connections between differential geometry and topology II," *Duke Mathematics Journal*, 2 (1936), 95–102.

17. Poincaré, H., "Sur les lignes géodésiques des surfaces convexes," *Transactions of the American Mathematics Society*, 6 (1905), 237–74.

18. Preismann, A., "Quelques propriétés globales des espaces de Riemann," *Comm. Math. Helv.*, 15 (1943), 175–216.

19. Rauch, H. E., "A contribution to differential geometry in the large," *Ann. of Mathematics*, 54 (1951), 38–55.

20. Rham, G. de, "Sur la réductibilité d'un espace de Riemann," *Comm. Math. Helv.*, 26 (1952), 328–44.

21. Schoenberg, J. M., "Some applications of the calculus of variations to Riemannian geometry," *Ann. of Mathematics*, 33 (1932), 485–95.

22. Struik, D. T., *Lectures on Classical Differential Geometry*. Reading, Massachusetts: Addison-Wesley Publishing Co., Inc., 1950.

23. Synge, J. L., "The first and second variations of the length in Riemannian spaces," *Proceedings of the London Mathematical Society*, 25 (1926), 247–64.

24. Topogonov, V. A., "On the convexity of Riemannian spaces with positive curvature," *Dokl. Nauk*, 115 (1957), 674–76.

25. Topogonov, V. A., "Riemannian spaces which have their curvature bounded from below by a positive number," *Dokl. Nauk*, 120 (1958), 719–21.

26. Topogonov, V. A., "Dependence between curvature and topological structure of Riemannian spaces of even dimensions," *Dokl. Nauk*, 133 (1960), 1031–33.

27. Tsukamoto, Y., "On Riemannian manifolds with positive curvature," *Memoirs of Fac. Sci. Kyushu Univ. Ser. A, Math.*, 15 (1962), 90–96.

28. Whitehead, J. H. C., "On the covering of a complete space by the geodesics through a point," *Ann. of Mathematics*, 36 (1935), 679–704.

Since the preparation of this manuscript, the following articles on the subject have appeared.

29. Bishop, R., and R. Crittenden, *Geometry of Manifolds*. New York: Academic Press Inc., 1964. This book is largely based on the lectures given by Ambrose, mentioned in the introduction.

30. Warner, F. W., "The conjugate locus of a Riemannian manifold," *American Journal of Mathematics*, 87 (1965), 575–604.

31. Warner, F. W., "Extensions of the Rauch comparison theorem to submanifolds," *Transactions of the American Mathematical Society*, 122 (1966), 341–56.

SURFACE AREA

Lamberto Cesari

There is a growing interest in a general theory concerning the analytical properties of transformations, or mappings

(1) $(T, A): p = p(w)$, $w \in A \subset X$, $p \in Y$, or $T: A \to Y$,

from a set A of a "space X" into a "space Y." Let us say explicitly that T is meant to be single-valued but not necessarily one-one—that is, each $w \in A$ is mapped into one and only one $p = p(w) \in Y$ (image of w), all these $p = p(w)$, $w \in A$, form a set $T(A) \subset Y$ (graph) of T, but each $p \in T(A)$ may be the image of more than one $w \in A$, even infinitely many $w \in A$ (counter images of p). Real functions of one real variable, parametric curves, surfaces, etc., are examples of such mappings, and the analytical entities attached to (T, A) may be called total variation, length, area, etc., or, more generally, line integrals, surface integrals, etc.

The last concepts are usually introduced under very restrictive conditions on T, but recently it has been recognized that length, area, etc., can be introduced under the mere hypothesis of continuity of T (and even this may not be required), and that the finiteness of the length (area, etc.) assures the existence of a line integral (surface integral, etc.).

If X and Y are euclidean spaces E_k and E_N, respectively, of k and N dimensions, and A is a nondegenerate interval of X, or any open set of X, then we may say that T is a parametric k-variety in E_N (a parametric curve for $k = 1$, a parametric surface for $k = 2$). The theory for $k = 1$ and $k = 2$ has reached some degree of completeness and has been extended in a number of directions in the last decades. Hopefully, a theory adequate in scope, depth, and generality for k-varieties will be developed for $k > 2$. Recent books (2, 11) are dedicated to the theory. A series of articles (3–5) in the *American Mathematical Monthly* illustrate briefly some of the results. The present exposition draws upon these articles; we shall discuss some of the questions concerning surfaces because surfaces entail a deeper connection with topology and measure theory than curves.

A completely abstract, or axiomatic approach has been initiated in Reference (8), where the underlying structure, or formal theory alone has been emphasized for general transformations. Nevertheless, we shall discuss here only actual continuous mappings (precisely surfaces) and their geometrical, analytical, and topological properties.

1. THE CONCEPT OF SURFACES

We shall denote by E_2 the real euclidean w-plane, $w = (u, v)$, by E_N any real euclidean space [for $N = 3$ let E_3 be the p-space with $p = (x, y, z)$], by $\overline{M}$, M^*, and M^0 the closure, the boundary, and the set of the interior points of a set M in any such space, by $|p| = (x^2 + y^2 + z^2)^{1/2}$ the euclidean norm, and by $|p - q|$ the euclidean distance of two points p, q.

By a surface S we shall mean a mapping (1) from some set $A \subset E_2$ into E_3 (or E_N), where A might be, for example, a square, a circle, a polygonal region, or a (closed) simple Jordan region J. Or, A could be, more generally, a (closed) Jordan region of finite connectivity $\nu \geqq 0$, say $J = J_0 - (J_1 + \cdots + J_\nu)^0$, $J_i \subset J_0^0$, $J_i J_j = 0, i \neq j, i = 1, \cdots, \nu$, where all J_i and J_0 are closed simple

Jordan regions. It has been found convenient to take for A any "admissible" set—that is, either any Jordan region J as above, or a finite sum of disjoint Jordan regions, or any set $G \subset E_2$, open in E_2, or any set $G \subset J$, open in J (further generalizations have been, and are being, studied). We will suppose $N = 3$. Thus (T, A) is defined by a continuous vector function $T(w)$, $w \in A$, say,

$$S = (T, A) : p = T(w), w \in A, T(w) = [x(w), y(w), z(w)], \text{ or}$$

(1.1)

$$S = (T, A) : x = x(u, v), y = y(u, v), z = z(u, v), (u, v) \in A.$$

The set $T(A) \subset E_3$ is the graph of $S = (T, A)$, but it in no way defines (T, A). For example, the mappings $T : x = u, y = v, z = 0$, $(u, v) \in Q$, and $T' : x = \sin^2 k\pi u$, $y = v$, $z = 0$, $(u, v) \in Q$, where Q is the square $Q = [0 \leqq u, v \leqq 1]$ and $k > 1$ an integer, have both the same graph $T(Q) = T'(Q) = U$, where U is the unit square $[0 \leqq x, y \leqq 1, z = 0]$ in the x, y-plane. However T covers U just once, while T' covers U exactly k times. Finally, if $P : x = \phi(t), y = \psi(t), 0 \leqq t \leqq 1$, denotes the well-known Peano curve covering U, then $T'' : x = \phi(u), y = \psi(u), z = 0, (u, v) \in Q$, has the same graph as T' and T, but is a completely different surface. Indeed, we shall mention that the "areas" of T, T', T'' are different numbers, namely 1, k, and 0, respectively.

Actually there are cases where our intuition associates to different mappings (or surfaces) the same entity (as, for example, T and T' with $k = 1$). Indeed, various concepts of "equivalence" have been taken into consideration, such as the Lebesgue and Fréchet equivalences (for example, T and T' above for $k = 1$ are Lebesgue equivalent). Classes of equivalent mappings are then denoted as Lebesgue surfaces and Fréchet surfaces. The definitions have been discussed in References (2) and (11), and for curves also in (4). We prefer not to mention these concepts and refer to the book (2), which in the following discussion will often be quoted as (SA) followed by indications of page and section.

Surfaces, or mappings $S = (T, A)$ for which each point $p \in T(A)$ has only one counterimage are said to be simple [and if T^{-1} is

continuous, then T is a homeomorphism between A and $T(A)$].
If for every $p \in T(A)$ the set $T^{-1}(p) \subset A$ is connected, then T is
said to be monotone; if $T^{-1}(p) \subset A$ is totally disconnected, then T
is said to be light. For example, surfaces of the type

$$T: x = u, \qquad y = v, \qquad z = z(u, v), \qquad (u, v) \in Q,$$

that is, $T: z = z(x, y), (x, y) \in Q$, are said, somewhat improperly,
to be "nonparametric surfaces," and they are certainly simple.
The mapping $T: x = u, y = v, z = (1 - u^2 - v^2)^{1/2}, (u, v) \in Q =$
$(u^2 + v^2 \leq 1)$, the graph of which is a "halfsphere," is nonparametric and simple. The mapping $T: x = 2(r - r^2)^{1/2} \cos \theta, y =$
$2(r - r^2)^{1/2} \sin \theta, z = 2r - 1$, where $(u, v) \in Q, r \cos \theta = u,$
$r \sin \theta = v$, the graph of which is the whole sphere $x^2 + y^2 + z^2 = 1$,
is monotone, but not light since the point $p = (0, 0, 1)$ is the image
of all points $(u, v) \in Q^*$. The mapping $T: x = 1 - u^2, y = u - u^3,$
$z = v, (u, v) \in A = (-2 \leq u \leq 2, 0 \leq v \leq 1)$, the graph of which
is a portion of a cylinder with generatrices parallel to the z-axis,
is light, but not monotone.

Of particular interest are the "flat surfaces" or "plane mappings"—that is, those mappings with a graph contained in a plane.
For example, if $\tau_r, r = 1, 2, 3$, denote the projections of E_3 onto
the yz, zx, xy coordinate planes, say E_{21}, E_{22}, E_{23}, then $T_r = \tau_r T,$
$r = 1, 2, 3$, are plane mappings, namely,

$$\begin{aligned} T_1: x = 0, \qquad y = y(u, v), \qquad z = z(u, v), \qquad (u, v) \in A, \\ (1.2) \quad T_2: x = x(u, v), \qquad y = 0, \qquad z = z(u, v), \qquad (u, v) \in A, \\ T_3: x = x(u, v), \qquad y = y(u, v), \qquad z = 0, \qquad (u, v) \in A. \end{aligned}$$

It was pointed out already by S. Banach and G. Vitali that properties of a mapping T (for example, the finiteness of the "area")
have no, or very little bearing on the properties of the single
functions $x(u, v), y(u, v), z(u, v)$, but they have an essential bearing
on the properties of the pairs of functions $(y, z), (z, x), (x, y)$, that
is, on the plane mappings $T_r, r = 1, 2, 3$. For example, no matter
which continuous function $\phi(u, v), (u, v) \in A$, we consider, the
mapping $T: x = y = z = \phi(u, v), (u, v) \subset A$, has a graph completely contained in the straight line $x = y = z$ in E_3, and its
"area" is zero. For nonparametric surfaces $T: x = u, y = v,$

$z = z(u, v)$, all properties of T are, of course, reflected into properties of the single function $z(u, v)$.

2. LEBESGUE AREA

For surfaces (mappings) $S = (T, A): p = p(w)$, $w \in A$, $p = (x, y, z)$, $w = (u, v)$, defined by functions $x(u, v)$, $y(u, v)$, $z(u, v)$, which are continuous in A with their first partial derivatives, it is usual to assume for the area of S the value of the integral (area integral)

$$(2.1) \quad I(T, A) = (A^0) \int |J| \, du \, dv = (A^0) \int (J_1^2 + J_2^2 + J_3^2)^{1/2} \, du \, dv,$$

and for the surface integral the value of

$$(2.2) \qquad I(T, A, f) = (A^0) \int f[p(w), J(w)] \, du \, dv,$$

where $J = J(w)$ is the vector Jacobian $J = (J_1, J_2, J_3)$, $J_1 = y_u z_v - z_u y_v$, etc., and where $f(p, t)$ is any given function of (p, t) continuous in (p, t) for all $p = (x, y, z) \in T(A)$, and all $t = (t_1, t_2, t_3)$. To assure that $I(T, A, f)$ is invariant with respect to Lebesgue or Fréchet equivalences, we add another condition

$$(h): f(p, kt) = kf(p, t) \quad \text{for all } k \geq 0, \quad t \in E_3 \text{ and } p \in T(A).$$

Thus, for $f = |t|$, $I(T, A, f)$ is the area integral $I(T, A)$ and condition (h) is satisfied. The definitions (2.1) and (2.2) are adequate under the restrictive conditions mentioned but inadequate generally. To see this, let us denote by $\phi(t)$, $0 \leq t \leq 1$, $\phi(0) = 0$, $\phi(1) = 1$, the well-known monotone nondecreasing function which is constant on each complementary interval I_j of the ternary Cantor set in $(0, 1)$. Thus, $\Sigma |I_j| = 1$ and $\phi'(t) = 0$ almost everywhere in $(0, 1)$ (9, p. 368). Then the nonparametric light mapping (surface) $S': x = u$, $y = v$, $z = \phi(u)$, $(u, v) \in Q = (0, 1, 0, 1)$ should have "area" greater than, or equal to, $\sqrt{2}$, while the integral $I(T, A)$ has the value 1. Analogously, the monotone mapping (surface) $S'': x = \phi(u), y = v, z = 0$ [a monotone mapping from Q into the square $U = [0 \leq x, y \leq 1, z = 0]$) should have "area" 1 (or at least 1), while the integral $I(T, A)$ has the value 0.

A definition shown to be adequate was proposed by Lebesgue in 1900. Simply, the Lebesgue area of a surface S is the "lower limit" of the elementary areas of the polyhedral surfaces approaching S. To make this definition precise, we need a few more words.

We shall denote as a figure F any finite sum of disjoint closed polygonal regions in E_2 (for example, a square, a polyhedral region). A mapping $(P, F): x = p(w)$, $w \in F$, from a figure F is said to be quasilinear, or a polyhedral surface if (1) $p(w)$ is single valued and continuous in F, and (2) there exists some subdivision D of F into nonoverlapping triangles t such that each component $x(u, v)$, $y(u, v)$, $z(u, v)$ of $p(w)$ is linear in each $t \in D$—that is, of the form $au + bv + c$, a, b, and c are constants for each t. Then P maps each $t \in D$ into a triangle $\Delta \subset E_3$, which may be degenerated into a segment, or a single point. Then by the elementary area of $a(P, F)$ of (P, F) is meant the sum $a(P, F) = \Sigma \, a(\Delta)$, where $a(\Delta)$ is the usual area of the triangle $\Delta = P(t)$ and Σ ranges over all $t \in D$. We shall say that a sequence $[A_n]$ of admissible sets A_n invades an admissible set A, if $A_n \subset A$, $A_n \subset A_{n+1}$, $A_n^0 \to A^0$ as $n \to \infty$. Finally, a sequence (T_n, A_n), $n = 1, 2, \cdots$, is said to be convergent toward (T, A) if (1) A_n invades A, and (2) $d_n \to 0$ as $n \to \infty$ where $d_n = \sup |T_n(w) - T(w)|$ for all $w \in A_n$. Thus, if $A_n = A$, $n = 1, 2, \cdots$, the convergence of (T_n, A) toward (T, A) is the uniform convergence in A of $T_n(w)$ toward $T(w)$, $w \in A$.

Suppose now that (T, A) is a given continuous mapping from an admissible set A and denote by γ the class of all sequences $[(P_n, F_n), n = 1, 2, \cdots]$ of quasilinear mappings convergent toward (T, A). Then the Lebesgue area $L(T, A)$ of (T, A) is defined by

$$(2.3) \qquad L(S) = L(T, A) = \operatorname*{Inf}_{\gamma} \liminf_{n \to \infty} a(P_n, F_n).$$

Of course some reader may ask why this definition is chosen instead of considering simply "the supremum, or the limit of the elementary areas of the polyhedral surfaces inscribed in S." H. Schwarz and G. Peano discovered in 1890 that such a supremum may be $+\infty$ and such a limit may not exist even with as simple a surface as a portion of "circular cylinder" (SA, 4.2, p. 24). Definition (2.3), proposed by Lebesgue in 1900, is the analogue of

one of the alternative definitions of Jordan length for a curve
(4). It can be proved (SA, 5.9, p. 37) that there exists some
sequence (P_n, F_n), $n = 1, 2, \cdots$, convergent toward (T, A) with
$\lim a_n(P_n, F_n) = L(T, A)$ as $n \to \infty$.

In case A is itself a figure (for example, a square) it is not restrictive to suppose (SA, 6.2, p. 61) that $F_n = A$ for all n. Then γ is
the class of all sequences $[(P_n, A), n = 1, 2, \cdots]$ of quasilinear
mappings convergent uniformly in A toward (T, A).

In case A is a figure and T is quasilinear, then $L(T, F) = a(T, F)$
—that is, the Lebesgue area coincides with the elementary area.
The proof of this fact, intuitive as it may appear, is difficult (see
SA, p. 108).

3. THE LOWER SEMICONTINUITY PROPERTY
OF LEBESGUE AREA

Definition (2.3) has, among others, the advantage that it makes
it easy to prove the lower semicontinuity of $L(S)$, which can be
expressed by the following statement.

THEOREM 3.1: *If (T_n, A_n), $n = 1, 2, \cdots$, is convergent toward*
(T, A), then $L(T, A) \leqq \lim \inf L(T_n, A_n)$ as $n \to \infty$.

Proof: We shall prove (3.1) in the case A is a given figure,
$A_n = A$ for all n, and by assuming in Definition (2.3) $F_n = A$ for
all n, according to the remark at the end of Section 2. We may
also assume $\lambda = \lim \inf L(T_n, A_n) < +\infty$, and $L(T_n, A) < +\infty$
for all n. Then, if $d_n = \max |T_n(w) - T(w)|$, we have $d_n \to 0$ as
$n \to +\infty$. By the definition of $L(T, A)$ there exists some sequence
(P_{nm}, A), $m = 1, 2, \cdots$, of quasilinear mappings convergent toward (T_n, A) as $m \to \infty$ with $a(P_{nm}, A) \to L(T_n, A)$. Thus we
have $\delta_{nm} \to 0$ as $m \to \infty$ where $\delta_{nm} = \max |P_{nm}(w) - T_n(w)|$ for
all $w \in A$. For each n, there exists an integer $m = m(n)$ such
that $\delta_{nm} < 1/n$, $|a(P_{nm}, A) - L(T_n, A)| < 1/n$. Now we have
$|P_{nm}(w) - T(w)| \leqq |P_{nm} - T_n| + |T_n - T| \leqq \delta_{nm} + d_n < d_n +$
$1/n$ for all $w \in A$. If $P'_n = P_{n,m(n)}$, then the sequence (P'_n, A), $n =$
$1, 2, \cdots$, converges toward (T, A), that is, belongs to the class γ
relative to (T, A). Thus, by Definition (2.3) we have

$$L(T, A) \leqq \liminf_{n \to \infty} a(P'_n, A) \leqq \liminf_{n \to \infty} \left[L(T_n, A) + \frac{1}{n} \right] = \lambda.$$

We shall denote as before by L and λ the numbers $L = L(T, A)$ and $\lambda = \liminf L(T_n, A_n)$. Even for polyhedral surfaces we may have $L < \lambda \leqq +\infty$ (in particular, we may have $L < +\infty$, $\lambda = +\infty$). We give the following example. Suppose $A = F$ is the unit square $A = (0 \leqq u \leqq 1, 0 \leqq v \leqq 1)$, and T is the identity mapping $T: x = u$, $y = v$, $z = 0$. Thus, $S = (T, A)$ is the unit square in the x, y-plane. Suppose $A_n = A$ and define $S_n = (T_n, A)$ as follows: $x_n = u$, $y_n = v$, $z_n = h_n(u - (i/n))$ for $i/n \leqq u \leqq (i/n) + (1/2n)$, $z_n = h_n((i/n) + (1/n) - u)$ for $(i/n) + (1/2n) \leqq u \leqq (i/n) + (1/n)$, where $i = 0, 1, \cdots, n - 1$, $0 \leqq u \leqq 1$, $0 \leqq v \leqq 1$ and $h_n, n = 1, 2, \cdots$, are arbitrary positive numbers. As n describes the sequence $n = 1, 2, \cdots$, we have a sequence of polyhedral surfaces $S_n, n = 1, 2, \cdots$. Each S_n consists of $2n$ rectangular strips of side lengths $\rho = 1$ and $\rho' = (1/2n)(1 + h_n^2)^{1/2}$. Each strip has a side of length 1 on the x, y-plane, and the opposite side is above and parallel to this plane at a height of $h_n/2n$. If we suppose all numbers h_n equal to 1, then the strips are all at 45-degree angles with the xy-plane, and we have $h_n/2n = 1/2n \to 0$ as $n \to \infty$; if we suppose $h_n = n^{1/2}$, the angle of the $2n$ strips with the xy-plane becomes closer and closer to 90° as n increases, but still we have $h_n/2n = 2^{-1}n^{-1/2} \to 0$ as $n \to \infty$. Because $x_n = x$, $y_n = y$, $|z_n - z| = h_n/2n$, we have $T_n \to T$ as $n \to \infty$ (uniformly on A). On the other hand, we have $L = L(T, A) = a(T, A) = 1$, and $L(T_n, A) = a(T_n, A) = (1 + h_n^2)^{1/2}$. Thus, for $h_n = 1$, $L(T_n, A) = 2^{1/2}$, $n = 1$, $2, \cdots$, and $L = 1$, $\lambda = 2^{1/2} > 1$. For $h_n = n^{1/2}$, $L(T_n, A) = (1 + n)^{1/2}$, and $L = 1$, $\lambda = +\infty$.

We shall return to the concept of area in Section 6. We mention here that an axiomatic approach to area has long been sought. For simplicity, suppose that A is a fixed figure F, and consider the class of all continuous mappings (T, A) from A into E_3. If a functional $Z(T, A)$ defined in this class must be considered an area, it should satisfy certain axioms, a number of which have been proposed. Of these, we list only a few.

1. $Z(T, A)$ *is lower semicontinuous, that is, it satisfies Theorem* (3.1) *(with L replaced by Z).*

2. $Z(P, A) = a(P, A)$ *for every quasilinear mapping* (P, A).

3. *There exists a sequence* (P_n, A) *of quasilinear mappings convergent toward* (T, A) *with* $a(P_n, A) \to Z(T, A)$ *as* $n \to \infty$.

Obviously the Lebesgue area satisfies all three axioms.

THEOREM 3.2: *Any functional* $Z(T, A)$ *satisfying Axioms* 1 *and* 2 *is less than, or equal to,* $L(T, A)$. *Thus, the Lebesgue area is the maximum of all functionals satisfying Axioms* 1 *and* 2.

Proof: Indeed, for every sequence (P_n, A) convergent toward (T, A) with $a(P_n, A) \to L(T, A)$, we have, by Axioms 1 and 2,

$$Z(T, A) \leqq \lim \inf Z(P_n, A) = \lim \inf a(P_n, A) = L(T, A),$$

that is, $Z \leqq L$, and Theorem 3.2 is proved.

THEOREM 3.3: *Any functional* $Z(T, A)$ *satisfying Axioms* 1, 2, *and* 3 *coincides with* $L(T, A)$.

Proof: Indeed, if (P_n, A) is the sequence mentioned in Axiom 3 for the mapping (T, A), we have, by Axioms 3 and 2, and by the definition of Lebesgue area,

$$Z(T, A) = \lim \inf Z(P_n, A) = \lim \inf a(P_n, A) \geqq L(T, A),$$

and thus $Z \geqq L$ and $Z \leqq L$—that is, $Z = L$. Thereby Theorem 3.3 is proved.

4. PLANE MAPPINGS: THEIR TOTAL VARIATION AND ABSOLUTE CONTINUITY

Let $(T, A):p = T(w)$, $w \in A$, $w = (u, v)$, $p = (x, y)$, be any continuous mapping from an admissible set A of the oriented u, v-plane E_2 into the oriented x, y-plane E_2', that is,

$$(T, A):x = x(u, v), \qquad y = y(u, v), \qquad (u, v) \in A.$$

For every simple closed polygonal region $\pi \subset A$, let us consider the oriented boundary π^* and the image $C:(T, \pi^*)$ of π^*—that is, the continuous mapping (T, π^*) defined by T on π^* (restriction of T on π^*). It is a closed oriented curve C of the x, y-plane E_2. For each point $p_0 = (x_0, y_0)$ not on the graph (C) of C, we can define the topological index $O(p_0; C)$ of C with respect to the point $p_0 \in E_2' - (C)$. Roughly speaking, $O(p_0; C)$ is the integral number of times ($\lessgtr 0$) in which C "links" the point p_0 in the positive direction. Suppose we assume on π^* a parameter, say s, $0 \leq s \leq a$. Suppose also we use polar coordinates (ρ, θ) of center p_0 in E_2. Then as s describes $(0, a)$—that is, w describes π^* once in the positive sense, $p = T(w)$ describes C. The modulus $\rho = \rho(s)$ of p—that is, $\rho = \rho(s) = |p - p_0|$, is a single-valued continuous function of s on $(0, a)$ and $\rho(a) = \rho(0)$. The argument $\theta(s)$ of p is defined only as mod 2π. If we fix any one of its values, say $\bar{\theta} = \theta(0)$, for $s = 0$, then by continuity, $\theta(s)$ is defined on $(0, a)$ as a single-valued continuous function of s, and $\theta(a) \equiv \theta(0) = \bar{\theta}$ (mod 2π). Then, by definition, $O(p_0; C) = (2\pi)^{-1}[\theta(a) - \theta(0)]$, and it is easy to prove that $O(p_0; C)$ does not depend on the parametrization of π^*, and on the choice of $\bar{\theta}$. This purely analytical definition (1, p. 462) of $O(p_0; C)$ is certainly the most elementary one. Other equivalent and purely topological definitions are given in topology. If C is rectifiable, and we think of E_2' as the plane of the complex variables $\zeta = x + iy$, with $\zeta_0 = x_0 + iy_0$, then $O(p_0; C)$ is given by the line integral

$$O(p_0; C) = (2\pi i)^{-1} \int_c (\zeta - \zeta_0)^{-1} \, d\zeta.$$

The topological index $O(p; C)$ has a number of analytical and topological properties (SA, 8.3–8.11, p. 85). Let us mention here that $O(p; C)$, $p = (x, y) \in E_2' - (C)$, is always finite (but not necessarily bounded in E_2'); $O(p; C)$ is constant on each complementary component of the graph (C) of C and is 0 in the unbounded one. If we assume $O(p, C) = 0$ at all points $p \in (C)$, then $O(p; C)$, $p \in E_2'$, is a single-valued, integral-valued ($\lessgtr 0$) function of p in E_2', and $O(p; C)$ is B-measurable. Thus, the L-integral

$$v(T; \pi) = (E_2') \int |O(p; C)| \, dx \, dy$$

exists (finite, or $+\infty$). Analogously, we may consider the numbers

$$v_+(T, \pi) = (E_2') \int O^+(p; C) \, dx \, dy \geqq 0,$$

$$v_-(T, \pi) = (E_2') \int O^-(p; C) \, dx \, dy \geqq 0,$$

where $O^+ = \frac{1}{2}[|O| + O]$, $O^- = \frac{1}{2}[|O| - O]$, and $v_+ + v_- = v$. Finally, if $O(p; C)$ is L-integrable, that is, $v(T, \pi) < +\infty$, then the number

$$u(T, \pi) = (E_2') \int O(p; C) \, dx \, dy \gtrless 0,$$

also exists, and $|u| \leqq v$.

If D denotes any finite system of closed, nonoverlapping, simple, polygonal regions $\pi \subset A$, let

$$V(T, A) = \sup_D \sum_{\pi \in D} v(T, \pi),$$

and for every point $p \in E_2'$,

$$N(p; T, A) = \sup_D \sum_{\pi \in D} |O(p; C)|.$$

Then $N(p; T, A)$, $0 \leqq N \leqq +\infty$, is an integral-valued, single-valued function of p in E_2', and $N(p)$ is lower semicontinuous in E_2—that is, $N(p_0) \leqq \liminf N(p)$ as $p \to p_0$, for every $p_0 \in E_2'$. The function $N(p, T, A)$, $p \in E_2'$, is said to be the multiplicity function of (T, A). [It is similar to the corrected multiplicity of a real function of one real variable considered in Reference 3 (p. 239).] Finally, let

$$W(T, A) = (E_2') \int N(p; T, A) \, dx \, dy.$$

Both V and W can be considered as total variations of (T, A). Indeed, V is of the Jordan total variation type, and W of the Banach total variation of a real function of one real variable type (3, p. 321). The following basic theorem extends the Banach theorem for real functions (3, p. 321).

THEOREM 4.1: *For every continuous plane mapping (T, A) we have $V(T, A) = W(T, A) = L(T, A)$.*

For a proof see (SA, pp. 186 and 390). Then the common value (finite, or $+\infty$) of V, W, L can be assumed as the total variation of the plane mapping (T, A), and (T, A) is said to be of bounded variation if $V(T, A) = W(T, A) < +\infty$. Analogously, let us write

$$V_+(T, A) = \sup_D \sum_{\pi \in D} v_+(T, \pi),$$

$$V_-(T, A) = \sup_D \sum_{\pi \in D} v_-(T, \pi),$$

$$N_+(p; T, A) = \sup_D \sum_{\pi \in D} O^+(p; C),$$

$$N_-(p; T, A) = \sup_D \sum_{\pi \in D} O^-(p; C),$$

$$W_+(T, A) = (E_2') \int N_+(p; T, A) \, dx \, dy,$$

$$W_-(T, A) = (E_2') \int N_-(p; T, A) \, dx \, dy.$$

THEOREM 4.2: *For every continuous plane mapping (T, A) we have* (SA, p. 187)

$$V_+(T, A) = W_+(T, A), \qquad V_-(T, A) = W_-(T, A),$$
$$V_+ + V_- = V, \qquad W_+ + W_- = W.$$

Thus, the common values $V_+ = W_+$, $V_- = W_-$ can be assumed as the positive and negative total variations of (T, A), and the difference $\mathfrak{B}(T, A) = V_+(T, A) - V_-(T, A)$ as the signed, or relative, total variation of (T, A).

Finally, if (T, A) is of bounded variation—that is, $V(T, A) < +\infty - v(T, \pi) < +\infty$ for every $\pi \in A$ and thus also is the following number defined:

$$U(T, A) = \sup_D \sum_{\pi \in D} |u(T, \pi)|.$$

THEOREM 4.3: *For every continuous bounded variation plane mapping (T, A) we have $U(T, A) = V(T, A)$.*

Let us observe finally that V is "overadditive"; that is, if (A') is any finite subdivision of A into nonoverlapping admissible sets, or, more generally, any finite system of nonoverlapping admissible sets $A' \subset A$, then $V(T, A) \geqq \sum V(T, A')$, and the sign $>$ may hold even in the apparently elementary case where A' and A are polygonal regions.

These considerations show how the concept of "plane mapping of bounded variation" can be founded on topological and measure theoretical considerations. On the same basis, we can introduce the corresponding concept of absolute continuity.

A continuous (plane) mapping $(T, A): p = T(w)$, $w \in A$, $w = (u, v)$, $p = (x, y)$, from the uv-plane E_2 into the xy-plane E_2', is said to be absolutely continuous if both the following conditions hold.

1. *Given* $\epsilon > 0$, *there is a* $\delta = \delta(T, A, \epsilon) > 0$ *such that for each finite system* $D = (\pi)$ *of nonoverlapping simple closed polygonal regions* $\pi \subset A$ *with* $\Sigma |\pi| < \delta$ *we have* $\Sigma v(T, \pi) < \epsilon$.

2. *For every simple closed polygonal region* $\pi \subset A$ *and finite subdivision* (π') *of* π *into nonoverlapping simple polygonal regions* π' *we have* $V(T, \pi) = \Sigma V(T, \pi')$.

In (1), $|\pi|$ denotes the Lebesgue measure of the set π in E_2 (area). Condition (1) is familiar, and essentially requires that v (and thus V) is "an absolutely continuous set function." Condition (2) simply requires that V is "additive" (at least in the class of the simple polygonal regions $\pi \subset A$). Conditions (1) and (2) are independent, as examples have shown, and it is merely their logical union (1) $\cup$ (2) which we assume here as a definition of absolute continuity. There are many interesting properties, each of which is a necessary and sufficient condition for a mapping (T, A) to be absolutely continuous.

5. A CHARACTERIZATION OF SURFACES
WITH FINITE AREA

The basic concepts of area of a mapping (T, A), $T: A \to E_3$ (Section 2), and of total variation of the plane mappings (T_r, A) defined by Equation (1.2) in Section 4, are connected by a basic theorem, which extends formally to Lebesgue area, a known Jordan theorem for Jordan length (4, p. 489).

THEOREM 5.1: *For every continuous mapping* $(T, A): p = T(w)$, $w \in A$, $w = (u, v)$, $p = (x, y, z)$, *we have*

(5.1) $V(T_r, A) \leqq L(T, A) \leqq V(T_1, A) + V(T_2, A) + V(T_3, A),$
$$r = 1, 2, 3.$$

*Thus, $L < +\infty$ if and only if all plane mappings (T_r, A), $r = 1, 2,$
3, are of bounded variation* (Cesari, 1942).

This theorem, despite its analogy with the elementary Jordan
theorem for curves [see (4)], has been shown to have a deep
topological and measure theoretical basis. The proof is given in
(SA, p. 295), and consists in the process of stretching and smoothing the continuous surface $S = (T, A)$ into continuous polyhedral
surfaces $S_n = (P_n, F_n)$, $n = 1, 2, \cdots$, with $P_n \to T$ as $n \to \infty$,
and $a(P_n, F_n) \leqq V_1 + V_2 + V_3$, $V_r = V(T_r, A)$, $r = 1, 2, 3$.

6. PEANO AND GEÖCZE AREAS

Let $(T, A): p = p(w)$, $w \in A$, be any continuous mapping from
$A \subset E_2$ into E_3, and let us denote by T_1, T_2, T_3 the plane mappings
of Equation (1.2) which are the projections $T_r = \tau_r T$ of T on the
(y, z), (z, x), (x, y) coordinate planes E_{21}, E_{22}, E_{23}. If π denotes any
simple polygonal region $\pi \subset A$, π^* the oriented boundary of π,
$C:(T, \pi^*)$, $C_r:(T_r, \pi^*)$, $r = 1, 2, 3$, the continuous oriented closed
curves which are the images of π^* under T and T_r. Thus, $(C) \in E_3$,
$(C_r) \subset E_{2r}$, and C_r is the "projection" of C on E_{2r}. According to
Section 4, we put

$$v_r = v(T_r, \pi) = (E_{2r}) \int |O(p; C_r)| \, dp, \qquad r = 1, 2, 3,$$

$$v = (v_1, v_2, v_3), \qquad |v| = (v_1^2 + v_2^2 + v_3^2)^{1/2},$$

$$V(T, A) = \sup_D \sum_{\pi \in D} |v(T, \pi)|,$$

where the supremum is taken with respect to all finite systems
D of nonoverlapping simple polygonal regions $\pi \subset A$. Thus,
$v(T_r, \pi) \leqq v(T, \pi)$, $V(T_r, A) \leqq V(T, A) \leqq +\infty$, $r = 1, 2, 3$. The
number V can be thought of as an "area" of the surface $S = (T, A)$.
A variant of this definition follows. Denote by α any plane in E_3,
τ_α the projections of E_3 onto α, $T_\alpha = \tau_\alpha T$ the projection of T on α,
$C_\alpha:(T_\alpha, \pi^*)$ the oriented continuous closed curve which is the image

of π^* under T_α, $(C_\alpha) \subset \alpha$. Then put $v(T_\alpha, \pi) = (\alpha) \int |O(p; C_\alpha)|\, dp$, $v^*(T, \pi) = \sup_\alpha v(T_\alpha, \pi)$, and

$$P(T, A) = \sup_D \sum_{\pi \in D} v^*(T, \pi) = \sup_D \sum_{\pi \in D} \sup_\alpha v(T_\alpha, \pi).$$

Also, $P(T, A)$, like $V(T, A)$ and $L(T, A)$, is an "area" of T, in the sense that all three definitions (just as many others) are precise mathematical formulations of concepts which all strongly appeal to our intuition as areas. In fact, the following theorem has been proved.

THEOREM 6.1: *For every continuous mapping (surface)* (T, A) *we have* $L(T, A) = V(T, A) = P(T, A)$ (C. B. Morrey, T. Rado, L. Cesari, J. Cecconi).

The common value of the three numbers L, V, P, is defined as the area of (T, A), and is usually called the Lebesgue area of (T, A). The present definitions of V and P are the final result of successive refinements of a concept first proposed by Peano in 1890. This process of successive refinements is necessary for us to reach the basic identity in Theorem 6.1, which did not hold for the previous somewhat crude concepts. A number of authors, among them Z. de Geöcze, S. Banach, L. Tonelli, R Caccioppoli, T. Rado, E. J. McShane, C. B. Morrey, J. Cecconi, and L. Cesari, have contributed to these refinements. By common agreement, the area V is usually designated as the Geöcze area and P as the Peano area of $S = (T, A)$, to honor two mathematicians who proved to have such deep insight into the forthcoming theory. For a direct proof of Theorem 6.1, see (SA, p. 390). If $V(T, A) < +\infty$, then $V_r(T, A) < +\infty$, $r = 1, 2, 3$, and for every $\pi \subset A$, the following numbers also exist

$$u_r = u(T_r, \pi) = (E_{2r}) \int O(p; C_r)\, dp, \qquad r = 1, 2, 3,$$

$$u = (u_1, u_2, u_3), \qquad |u| = (u_1^2 + u_2^2 + u_3^2)^{1/2},$$

$$U(T, A) = \sup_D \sum_{\pi \in D} |u(T, \pi)|,$$

where the preceding conventions are used. Then we have, obviously,

$$0 \leqq |u(T, \pi)| \leqq v(T, \pi) < +\infty,$$

$$0 \leqq U(T, A) \leqq V(T, A) \leqq +\infty,$$

and it is clear that for some π and T, we may well have $|u| < |v|$. Nevertheless, the following theorem holds.

THEOREM 6.2: *For every mapping* (T, A) *with* $V(T, A) < +\infty$, *we have* $U(T, A) = V(T, A)$ (Cesari).

7. A WEIERSTRASS-TYPE INTEGRAL

By the same blending of topological and analytical considerations, by which we have defined Geöcze and Peano areas, we may now define the concept of surface integral through a Weierstrass-type limit process.

Let $S = (T, A): p = T(w)$, $w \in A$, $w = (u, v)$, $p = (x, y, z)$, be a given mapping of finite area, and, for simplicity, let us suppose that A is a closed finitely connected Jordan region. Let $f(p, t)$ be a continuous function of (p, t), $p = (x, y, z)$, $t = (t_1, t_2, t_3)$, for all $p \in T(A)$ and real vector t satisfying the usual condition $f(p, kt) = kf(p, t)$ for all $k \geqq 0, p \in T(A)$, and t. For every simple polygonal region $\pi \subset A$ let us consider the vector

$$u = u(T, \pi) = (u_1, u_2, u_3), \qquad u_r = u(T_r, \pi), \qquad r = 1, 2, 3,$$

of norm $|u| = (u_1^2 + u_2^2 + u_3^2)^{1/2}$. If $|u| > 0$, then the unit vector

$$a = a(T, \pi) = (a_1, a_2, a_3), \qquad a_r = \frac{u_r}{|u|},$$

$$r = 1, 2, 3, \qquad a_1^2 + a_2^2 + a_3^2 = 1$$

can be thought of as the vector of the direction cosines of an "average normal" n to the piece of the surface S defined by T on π. If D is any finite system $D = (\pi)$ of nonoverlapping simple polygonal regions $\pi \subset A$, and for each $\pi \in D$, we denote by $\bar{p}$ any point of $T(\pi)$, we may consider the sum

$$S = \sum_{\pi \in D} f(\bar{p}, u) = \sum_{\pi \in D} f(\bar{p}, a)|u(\pi, T)|$$

as an "approximate expression" of a Weierstrass-type integral $E(T, A, f)$ of f on the surface $S = (T, f)$. An index $\delta = \delta(D)$

measuring the "fineness" of D can be introduced in various ways. Then it can be proved that the following limit exists and is finite

$$E(T, A, f) = \lim_{\delta \to 0} S = \lim_{\delta \to 0} \sum_{\pi \in D} f(\bar{p}, a)|u(\pi, T)|,$$

for every continuous mapping (T, A) of finite area (Cesari). The integral E is invariant with respect to both Lebesgue and Fréchet equivalences; that is, $E(T, A, f)$ has the same value in correspondence with Lebesgue or Fréchet equivalent mappings.

An index δ of fineness of a system D is, for example, defined by the maximum of all following numbers: diam $T(\pi)$ for $\pi \in D$; $|\Sigma\, T_r(\pi)|$, $r = 1, 2, 3$, and

$$U(T, A) - \sum_{\pi \in D} |u(T, \pi)|,$$

$$U(T_r, A) - \sum_{\pi \in D} |u(T_r, \pi)|, \qquad r = 1, 2, 3.$$

8. RELATION BETWEEN AREA
AND AREA INTEGRAL

What relation exists between area and the area integral of Section 2, and between the Weierstrass-type integral E and the classical surface integral I of Section 2? As mentioned in Section 2, the finiteness of the area does not imply the existence of the partial derivatives of $x(u, v)$, $y(u, v)$, $z(u, v)$, and, hence, of the ordinary Jacobians $J_1 = y_u z_v - y_v z_u$, $\cdots$. The example $S = (T, A){:}x = y = z = \phi(u, v)$, where $\phi(u, v)$ is any continuous function with partial derivatives at no point, is typical, because $L(S) = 0$ is certainly finite here. Thus, it is clear that $L(S) < +\infty$ implies the existence of the Weierstrass type surface integral $E(T, A, f)$, but neither area integral $I(T, A)$ nor the classical integral $I(T, A, f)$ need exist.

Nevertheless, under the sole hypothesis of finiteness of the area, certain "generalized Jacobians" $\mathfrak{J}_r(w)$, $w \in A^0$, can be defined almost everywhere in A^0, and for which we have $\mathfrak{J}_r(w) = J_r(w)$, $r = 1, 2, 3$, almost everywhere in A^0, whenever the functions x, y, z have ordinary partial derivatives almost everywhere in A^0. Of the various equivalent definitions of generalized Jacobians, the following is certainly the simplest and holds for almost all $w_0 \in A^0$:

$$\mathfrak{I}_r(w_0) = \lim_{\sigma \to 0} \frac{\mathfrak{B}(T_r, q)}{|q|}, \quad r = 1, 2, 3,$$

where $w_0 \in A^0$, q denotes any square with sides parallel to the u and v axes, with $w_0 \in q$, $q \subset A^0$, and $\sigma = \operatorname{diam} q$. With these definitions, we can prove the following theorem, which corresponds formally to a Tonelli's theorem for curves (4, p. 492).

THEOREM 8.1: *For any continuous mapping* $S = (T, A)$ *with* $L(T, A) < +\infty$, *we have*

(8.1) $$L(T, A) \geqq (A^0) \int |\mathfrak{I}| \, du \, dv,$$

where $\mathfrak{I} = (\mathfrak{I}_1, \mathfrak{I}_2, \mathfrak{I}_3)$. *The equality sign holds if and only if the plane mappings* (T_r, A), $r = 1, 2, 3$, *are absolutely continuous.*

Suppose now that $S = (T, A)$ and $f(p, t)$ are given as in Section 7. Then we have the following theorem.

THEOREM 8.2: *If* $L(S) < +\infty$ *and the plane mappings* (T_r, A), $r = 1, 2, 3$, *are absolutely continuous, then*

(8.2) $$E(T, A, f) = (A^0) \int f[p(w), \mathfrak{I}(w)] \, du \, dv.$$

In other words, the integral E is given as an ordinary surface integral (with generalized Jacobians) whenever the area is given by the corresponding area integral.

In both Theorems 8.1 and 8.2 the integrals are L-integrals.

The question finally presents itself, whether any continuous surface $S = (T, A):x = x(u, v), y = y(u, v), z = z(u, v), (u, v) \in A$, has a "representation" $(T^*, A):x = X(u, v), y = Y(u, v), z = Z(u, v), (u, v) \in A$, for which the partial derivatives $X_u, \cdots, Z_v$ exist almost everywhere in A^0, and for which the area is given by the classical area integral. In other words, we ask whether another continuous mapping (T^*, A) exists which is Lebesgue, or Fréchet equivalent to (T, A), for which the plane mappings (T_r^*, A), $r = 1, 2, 3$, are absolutely continuous and for which $X_u, \cdots, Z_v$ exist almost everywhere in A^0. The answer is affirmative when Fréchet equivalence is considered (Cesari).

On the other hand, it was recently proved from an abstract and more general viewpoint [Turner, Cesari; see (8)] that for any continuous Fréchet surface S of finite Lebesgue area and any representation (T, A) of S [that is, for any element (T, A) of the equivalence class defining S], there is a suitable measure function μ (defined in a suitable algebra $\mathcal{C}$ of subsets M of A) such that

$$L(T, A) = (A^0) \int d\mu$$

$$E(T, A, f) = (A^0) \int f[p(w), \theta(w)] \, d\mu,$$

where $\theta(w)$ is a unit vector, $|\theta(w)| = 1$, defined μ almost everywhere in A^0 and representing a generalized normal vector to the surface S. In other words, every representation (T, A) of a surface S of finite Lebesgue area can be used for the computation of the area of S and of the surface integrals on S, provided we express these as Lebesgue-Stieltjes integrals in terms of a suitable measure function (area measure), and a suitable generalized normal vector function. As stated previously, we refer to (8) for this more general viewpoint, and we shall now return to the analytical and topological properties of continuous surfaces.

9. FINE-CYCLIC ELEMENTS

For simplicity, we shall suppose, as in Section 7, that A is a closed, finitely connected, Jordan region, say $A = J_0 - J_1 + \cdots + J_\nu)^0$ (Section 1), where $0 \leq \nu \leq +\infty$ is the order of connectivity of A. Thus, for $\nu = 0$, A is called a disc; for $\nu = 1$, A is called an annular region. We have already mentioned in Section 1 the concepts of monotone mappings and of light mappings. We should now recall a precise characterization of the topological structure of any continuous mapping. For the purpose, let us mention here that if we have a mapping $f: y = f(x)$ from a "set" A onto a set B, and a mapping $g: z = g(y)$ from B into a space C, then the composition mapping, $F = gf: z = g[f(x)]$ from A into C, is said to be the product of f and g (in this order) and denoted by gf. Also, $F = gf$ is said to be a factorization of F into the two factors f and g. With this convention, we may state at once the factorization theorem of analytic topology, namely, that every continuous mapping, say $(T, A): p = p(w), w \in A$, has a factorization $T = lm$ into two factors, a monotone mapping m, followed by a light mapping l (monotone-light factorization).

To understand this statement in the terms necessary here, let us consider for every point $p \in T(A)$ the inverse set $T^{-1}(p) \subset A$. Since A is compact and T is continuous, the set $T^{-1}(p)$ is closed; hence, its components g are continua $g \subset A$. By a "continuum" is meant, as usual, a bounded, closed, connected set, and hence even single points may be considered continua. Note that, if T is monotone (Section 1), then, for every $p \in T(A)$, the inverse set $g = T^{-1}(p) \subset A$ is just one continuum; if T is light (Section 1), then, for every $p \in T(A)$, all components g of $T^{-1}(p) \subset A$ are single points. In any case, for any continuous mapping (T, A), the collection $\Gamma = \{g\}$ of all components g of the set $T^{-1}(p)$, $p \in T(A)$, is a decomposition of A into disjoint continua $g \subset A$ (which may well be all single points of A). Γ is the collection of all maximal continua of constancy for T in A.

We may also consider the family $\mathcal{G}_0$ of all sets $G \subset A$ which are open in A or are the union of continua $g \in \Gamma$, say $G = \cup g$ (that is, have the property that $g \in \Gamma$; $gG \neq 0$ implies $g \subset G$).

We may now consider the elements g of Γ as "points," say $\tilde{g}$, and Γ as the "space," say $\tilde{\Gamma}$ the points of which are $\tilde{g}$. To consider $\tilde{\Gamma}$ as a space, we actually should define a topology on $\tilde{\Gamma}$, which turns out to be the equivalent of defining in $\tilde{\Gamma}$ the collection $\tilde{\mathcal{G}}$ of the open sets $\tilde{G}$. We can do so easily by considering each set $G = \cup g$, $G \in \mathcal{G}_0$, as the union $\tilde{G} = \cup \tilde{g}$ of the corresponding elements $\tilde{g}$. Then $\tilde{\mathcal{G}}$ is simply the family of all sets $\tilde{G}$.

By these natural definitions, $\tilde{\Gamma}$ can be proved to be not only a topological space, but also a "Peano space" (in particular, compact, connected, locally connected). Now, if $m: \tilde{g} = m(w), w \in A$, is the mapping from A onto $\tilde{\Gamma}$, which maps each point $w \in g$, $g \in \Gamma$ into the point $\tilde{g} \in \tilde{\Gamma}$, then m is obviously monotone, because for each $g \in \tilde{\Gamma}$, the set $m^{-1}(\tilde{g}) = g$ is exactly the continuum $g \in \Gamma$, $g \subset A$. Finally, if $l: p = p(\tilde{g})$, $\tilde{g} \in \tilde{\Gamma}$, is the mapping from $\tilde{\Gamma}$ onto $T(A) \subset E_3$ defined by $l = Tm^{-1}$, then l is "light," because for each $p \in T(A)$, the components g of the set $l^{-1}(p)$ are all single points $\tilde{g} \in \tilde{\Gamma}$. While we refer to the usual expositions for more formal proofs, we note that $T = lm$ is a monotone-light decomposition of T, that $m:A \to \tilde{\Gamma}$, $l:\tilde{\Gamma} \to T(A)$, and that $\tilde{\Gamma} = m(A)$ is the middle space, or hyperspace, of the decomposition (15).

Obviously any "space" M which is homeomorphic to $\tilde{\Gamma}$ may be considered a middle space, or hyperspace, for T, because, if h is a homeomorphism of $\tilde{\Gamma}$ onto M, and $m' = hm$, $l' = lh^{-1}$, then $T = l'm'$, $m':A \to M$, $l':M \to T(A)$, and m' is monotone and l' is light. Thus, M is the middle space and can be called a model of $\tilde{\Gamma}$. Also, for every monotone-light factorization $T = lm$ of T, $m:A \to M$, $l:M \to T(A)$, M is homeomorphic to—that is, M is a model of $\tilde{\Gamma}$ (15).

If for some readers the previous considerations appear somewhat abstract, the following remark may be of help: A model M can be built in the euclidean space E_3. If T is light, we may take for m the identity mapping, and $M - A$ is the middle space; if T is monotone, we may take for l the identity mapping, and $M = T(A)$ —that is, the graph of (T, A) is the middle space M.

For example, in the monotone mapping $T:x = u$, $y = 0$, $z = 0$, $(u, v) \in A = (0 \leqq u, v \leqq 1)$, M is a segment (an arc); in the monotone mapping $T:x = \sin \pi r \cos \theta$, $y = \sin \pi r \sin \theta$, $z = \cos \pi r$, $(u, v) \in A = (u^2 + v^2 \leqq 1)$, where $r \cos \theta = u$, $r \sin \theta = v$, $M = T(A)$ is the unit sphere in E_3; in the light mapping $T:x = u$, $y = v$, $z = z(u, v)$, $(u, v) \in A$, where $z(u, v)$ is a continuous function constant on no proper subcontinuum of A, the middle space is $M = A$.

Let $(T, A):p = p(w)$, $w \in A = (u^2 + v^2 \leqq 1)$ be the monotone mapping from the disc A into E_3 defined by $x = (2r - 1) \cos \theta$, $y = (2r - 1) \sin \theta$, $z = 0$, if $\frac{1}{2} \leqq r \leqq 1$, and by $x = y = 0$, $z = 1 - 2r$, if $0 \leqq r \leqq \frac{1}{2}$, where $r \cos \theta = u$, $r \sin \theta = v$. Then Γ is the collection of all circles $u^2 + v^2 = r^2$ with $0 < r \leqq \frac{1}{2}$, and of all single points $(u, v) \in A$, with $u = v = 0$, or $\frac{1}{4} < u^2 + v^2 \leqq 1$. The middle space $M = T(A)$ is made up of the unit circle of the x, y-plane and of the unit segment of the z-axis issuing from its center (a disc and a thread).

Let us consider now mappings $(T, A):p = T(w)$, $w \in A = (0 \leqq u, v \leqq 1)$, from the unit square into E_3, where either Γ is the collection of all segments $g = (0 \leqq v \leqq 1, u = t)$, $0 \leqq t \leqq 1$ [surfaces $S_1 = (T, A)$], or Γ is the collection of the boundaries $g =$ [max ($|2u - 1|$, $|2v - 1|$) $= t$], $0 \leq t \leq 1$, of all squares contained in A, concentric and similar to A [surfaces $S_2 = (T, A)$]. In either case, we may assume $T(w) = f(t)$, where $f(t)$ is a continuous function of t in $0 \leq t \leq 1$, constant on no subinterval of $(0, 1)$. If we suppose that $f(t)$ never takes twice the same value $f = (x, y, z)$, then T is monotone, and $M = T(A)$ is an arc PQ. The two types of surfaces, S_1 and S_2, apparently identical in E_3, are different. Surfaces S_1 can be thought of as limit cases of thin strips, and surfaces S_2, as limit cases of thin cones. The two surfaces S_1 and S_2, even defined by the same vector function f, are neither Lebesgue nor Fréchet equivalent.

Mappings (T, A) with $L(T, A) = 0$ can be characterized topologically—namely, M is a space of dimension less than, or equal to, 1 (T. Rado, 1945, for $\nu = 0$; R. F. Williams, 1958, for $\nu \geq 0$). In the case where A is a disc ($\nu = 0$), M has been further characterized (as a dendrite of analytic topology). Simply stated, we may expect M to be a ramified system of threads.

Mappings $(T, A): p = T(w)$, $w \in A$, with $L(T, A) > 0$, must therefore possess a middle space M with some two-dimensional parts (see, for example, the preceding example where M consists of a disc and a thread). These parts are important and, where A is a disc ($\nu = 0$), they are the cyclic elements Σ of M. A subset Σ of M is said to be a cyclic element of M if Σ is a proper continuum and is not disconnected by suppressing any one of the points of Σ. Any two cyclic elements Σ of M are not overlapping (they may have in common at most one point), and the collection $\{\Sigma\}$ of all cyclic elements of M is at most countable. Thus, in the preceding example, the only cyclic element Σ of M is the disc. Again, for $\nu = 0$, the cyclic elements Σ of M have been fully characterized: each is either a disc or a sphere (15).

For $\nu > 0$, a further decomposition of M may be necessary. The parts σ of M of dimension 2 may actually be finer than the cyclic elements Σ of M, and they are denoted as the fine-cyclic elements σ of M. A subset σ of M is said to be a fine-cyclic element of M if σ is a proper continuum and is not disconnected by suppressing any finite system of points of σ (7 and 10). Any two fine-cyclic elements

σ of M are not overlapping (they may have in common at most finitely many points), and the collection $\{\sigma\}$ of all fine-cyclic elements σ of M is at most countable. For example, suppose that $\nu = 1$, A is the annular region $1 \leq u^2 + v^2 \leq 4$, and (T, A) is the monotone mapping defined by $x = \varphi(u)$, $y = v$, $z = 0$, where $\varphi(u) = u$ for $-1 \leq u \leq 1$, $= 1$ for $u \geq 1$, $= -1$ for $u \leq -1$. Then, the surface $S = (T, A)$ is made up of the two Jordan regions, $R_1 = [1 \leq x^2 + y^2 \leq 4, -1 \leq x \leq 1, y \geq 0]$ and $R_2 = [1 \leq x^2 + y^2 \leq 4, -1 \leq x \leq 1, y \leq 0]$, having in common the two points $(-1, 0)$ and $(1, 0)$. We may take $M = T(A) = R_1 + R_2$, and M presents two fine cyclic elements, R_1 and R_2, but only one cyclic element, M itself.

BIBLIOGRAPHY

On the general topic of this article, the reader may consult the books (2) and (11) and the articles (3–6) and (12), as well as the other books and papers listed in the bibliography.

1. Alexandroff, P., and H. Hopf, *Topologie.* Berlin: 1935.

2. Cesari, L., *Surface Area.* Princeton, N.J.: Princeton University Press, 1956 (quoted in article as *SA*).

3. Cesari, L., "Variation, multiplicity, and semicontinuity," *American Mathematical Monthly*, 65 (1958), pp. 317–32.

4. Cesari, L., "Rectifiable curves and the Weierstrass integral," *American Mathematical Monthly*, 65 (1958), pp. 485–500.

5. Cesari, L., "Recent results in surface area theory," *American Mathematical Monthly*, 66 (1959), pp. 173–92.

6. Cesari, L., "Retraction, homotopy, integral," address at the *International Congress of Mathematics*, Amsterdam, 1954.

7. Cesari, L., "Fine-cyclic elements of surfaces of type ν," *Rivista Mat. Univ. Parma*, 7 (1956), pp. 149–85.

8. Cesari, L., (a) "Quasi additive set functions and the concept of integral over a variety," *Transactions of the American Mathematical Society*, 102 (1962), pp. 94–113. (b) "Extension problem for quasi additive set functions and Radon-Nikodym derivatives," *Transactions of the American Mathematical Society*, 102 (1962), pp. 114–46.

9. Hobson, E. W., *The Theory of Functions of a Real Variable*, vol. I. Cambridge: Cambridge University Press, 1927.

10. Neugebauer, C. J., *B*-sets and fine-cyclic elements, *Transactions of the American Mathematical Society*, 88 (1958), pp. 121–36.

11. Rado, T., *Length and Area*. Providence: American Mathematical Society Colloquium Publication, vol. 30, 1948.

12. Silverman, E., "A miniature theory for Lebesgue area," *American Mathematical Monthly*, 67 (1960), pp. 424–30.

13. Williams, R. F., "Lebesgue area zero, dimension and fine-cyclic elements," *Rivista Mat. Univ. Parma*, 10 (1959), pp. 131–43.

14. Williams, R. F., "Lebesgue area of maps from Hausdorff spaces," *Acta Mathematica*, 102 (1959), pp. 33–46.

15. Whyburn, G. T., *Analytic Topology*. Providence: American Mathematical Society Colloquium Publication, vol. 28, 1942.

INTEGRAL GEOMETRY

L. A. Santalo

1. INTRODUCTION

We shall begin with three simple examples which will show the basic ideas on which integral geometry has been developed.

1.1. *Sets of points.* Let X be a set of points in the euclidean plane E_2. The measure (ordinary area) of X is defined by the integral

$$(1.1) \qquad m(X) = \int_X dx\, dy.$$

Let $\mathfrak{M}$ be the group of motions in E_2. With respect to an orthogonal Cartesian system of coordinates, the equations of a motion $u \in \mathfrak{M}$ are

$$(1.2) \qquad \begin{aligned} x' &= x \cos \varphi - y \sin \varphi + a \\ y' &= x \sin \varphi + y \cos \varphi + b. \end{aligned}$$

The fundamental property of the measure (1.1) is that of being invariant under $\mathfrak{M}$. That is, if $X' = uX$ is the transform of X by u, we have

$$(1.3) \qquad m(X') = \int_{X'} dx' \, dy' = \int_X dx \, dy = m(X)$$

as follows immediately from (1.2). It is well known that this propperty characterizes the measure (1.1) up to a constant factor.

Because we are generally interested only in the differential form under the integral sign in (1.1), we shall write $dP = dx \, dy$, or, more precisely,

$$(1.4) \qquad dP = dx \wedge dy$$

to indicate that the differential form under a multiple integral sign is an exterior differential form [see, for example, Munroe (43)].

The exterior differential form (1.4) is called the density for points in E_2 with respect to $\mathfrak{M}$. We shall always take the densities in absolute value.

1.2. *Sets of lines.* Let X now be a set of lines in E_2—for example, the set of all lines G which intersect a given convex domain K. We ask for a measure of X invariant under $\mathfrak{M}$.

Let p be the distance from the origin O to G and θ the angle formed by the perpendicular to G through O and the x-axis. We maintain that this invariant measure is given by

$$(1.5) \qquad m(X) = \int_X dp \, d\theta.$$

For a proof, we observe that by the motion u [Relation (1.2)] the line coordinates p, θ transform according to

$$(1.6) \quad \theta' = \theta + \varphi, \qquad p' = p + a \cos(\theta + \varphi) + b \sin(\theta + \varphi)$$

and putting $X' = uX$, we have

$$m(X') = \int_{X'} dp' \, d\theta' = \int_X dp \, d\theta = m(X)$$

which proves the invariance of $m(X)$. That this measure is unique, up to a constant factor, follows from the transitivity of the lines under $\mathfrak{M}$, since if $\int_X f(p, \theta) \, dp \, d\theta$ is invariant we must have $\int_{X'} f(p', \theta') \, dp' \, d\theta' = \int_X f(p, \theta) \, dp \, d\theta$, and, on the other hand, according to(1.6), $\int_{X'} f(p', \theta') \, dp' \, d\theta' = \int_X f(p', \theta') \, dp \, d\theta$. From the last two equalities, we obtain $\int_X f(p', \theta') \, dp \, d\theta = \int_X f(p, \theta) \, dp \, d\theta$. If this equality holds for any set X it must be true that $f(p', \theta') = f(p, \theta)$, and, since any line $G(p, \theta)$ can be transformed into any other $G(p', \theta')$ by a motion, we deduce $f(p, \theta) = $ constant.

The differential form

$$(1.7) \qquad\qquad dG = dp \wedge d\theta,$$

taken in absolute value, is called the density for lines in E_2 with respect to $\mathfrak{M}$.

Let us consider a simple application. To get the measure of the set of lines which cut a fixed segment S of length l, because of the invariance under $\mathfrak{M}$ we may take the origin of coordinates coincident with the middle point of S and the x axis coincident with the direction of S; then we have

$$(1.8) \quad m(G; G \cap S \neq 0) = \int_{G \cap S \neq 0} dp \, d\theta = \int_0^{2\pi} \left| \frac{l}{2} \cos \theta \right| d\theta = 2l.$$

If instead of S we consider a polygonal line Γ composed of a finite number of segments S_i of lengths l_i, writing (1.8) for each S_i and summing we get

$$(1.9) \qquad\qquad \int n \, dG = 2L$$

where $n = n(G)$ is the number of points in which $G(p, \theta)$ cuts Γ and L is the length of Γ. The integral in (1.9) is extended over all lines of the plane, n being 0 if $G \cap \Gamma = 0$. By a limit process it is not difficult to prove that (1.9) holds for any rectifiable curve [Blaschke (3)].

Conversely, given a continuum of points Γ in the plane, if the integral on the left of (1.9) has a meaning, then it can be taken as a definition for the length of Γ, which is the so-called Favard length [Nöbeling (45)].

For a convex curve K we have $n = 2$ for all G which intersect K, except for the positions in which G is a supporting line of K, which are of zero measure. Consequently we have: The measure

of the set of lines which intersect a convex curve is equal to its
length.

1.3. *Kinematic density.* Let us now consider a set X of oriented
congruent segments S of length l—for example, the set of those
which intersect a fixed convex domain. The position of S in E_2
is determined by the coordinates of its origin $P(x, y)$ and the
angle α formed by S and the x-axis. If we want to define a measure
for X invariant under $\mathfrak{M}$, we must take

$$(1.10) \qquad m(X) = \int_X dx \, dy \, d\alpha.$$

To see this, we first observe that by a motion (1.2) the variables
(x, y, α) transform according to (1.2) and $\alpha' = \alpha + \varphi$. Consequently the Jacobian of the transformation is 1, and we have

$$m(X') = \int_{X'} dx' \, dy' \, d\alpha' = \int_X dx \, dy \, d\alpha = m(X)$$

where $X' = uX$, which proves the invariance of $m(X)$. The
uniqueness, up to a constant factor, follows from the transitivity
of $\mathfrak{M}$ with respect to the congruent segments of the plane by the
same argument previously given for the lines.

If instead of segments we want to measure sets of congruent
figures K, since the position of such a figure is determined by the
position of a point $P(x, y)$ rigidly bound to K and the angle α
between a fixed direction PA in K and the x-axis, we can take
the same integral (1.10). The differential form

$$(1.11) \qquad dK = dx \wedge dy \wedge d\alpha$$

is called the kinematic density for E_2 with respect to the group $\mathfrak{M}$.
It is always taken in absolute value.

Another form for dK is obtained if instead of the coordinates
(x, y, α) for the oriented segment S, we take the coordinates
(p, θ) of the line G which contains S and the distance $t = HP$
from P to the foot H of the perpendicular drawn from the origin 0
to G. The transformation formulas are

$$(1.12) \quad x = p \cos \theta + t \sin \theta, \quad y = p \sin \theta - t \cos \theta, \quad \alpha = \theta - \frac{\pi}{2}$$

and consequently, up to the sign, we have $dx \wedge dy \wedge d\alpha = dp \wedge d\theta \wedge dt$. We may then write

$$(1.13) \qquad\qquad dK = \vec{dG} \wedge dt$$

where we write $\vec{G}$ in order to indicate that G must be considered as oriented ($\vec{dG} = 2\,dG$).

From this expression for dK we easily deduce the measure of the set of segments of length l which intersect a given convex domain K of area F and perimeter L. In fact, calling λ the length of the chord determined by G on K, we have

$$m(S; S \cap K \neq 0) = 2\int dp\,d\theta\,dt = 2\int_{S\|K \neq 0} (\lambda + l)\,dp\,d\theta$$
$$= 2\pi F + 2lL,$$

This formula can be generalized to surfaces [see (55)]; an application was given by Green (22).

If we ask for the measure of the set of segments S which are contained in K, the result is not simple; it depends largely on K. For instance, for a circle C of diameter $D \geqq l$, we have

$$m(S; S \subset C) = \frac{\pi}{2}\left(\pi D^2 - 2D^2 \text{ arc sin } \frac{l}{D} - 2l\sqrt{D^2 - l^2}\right)$$

and for a rectangle R of sides a, b ($a \geqq l$, $b \geqq l$), we have

$$m(S; S \subset R) = 2(\pi ab - 2(a + b)l + l^2).$$

An unsolved problem is that of finding among all convex domains K with a given perimeter those which maximize the measure $m(S; S \subset K)$ of the segments of a given length which are contained in K. For $l = 0$ the problem is the classical isoperimetric problem and the solution is well-known to be the circle.

The preceding very simple examples show the three steps which constitute the so-called integral geometry in the original sense of Blaschke (3): (1) definition of a measure for sets of geometric objects with certain properties of invariance; (2) evaluation of this measure for some particular sets; and (3) application of the obtained result to get some statements of geometrical interest.

The same examples show the basic elements which are necessary

to build the integral geometry from a general point of view:
(1) a base space E in which the objects we consider are imbedded
(in the preceding examples, E was the euclidean plane E_2); (2) a
group of transformations $\mathfrak{G}$ operating on E (in the preceding
examples $\mathfrak{G}$ was $\mathfrak{M}$; (3) geometric objects F contained in E which
transform transitively by $\mathfrak{G}$ (in the preceding examples, the geometric objects were points, lines or congruent figures).

Given E, $\mathfrak{G}$, and F, the first problem of the integral geometry
is to find a measure for sets of F invariant under $\mathfrak{G}$.

2. GENERAL INTEGRAL GEOMETRY

2.1. *Density and measure for groups of matrices.* Though the
integral geometry deals with general Lie groups, from the geometrical point of view in which we are principally interested it
suffices to consider Lie groups which admit a faithful representation, that is, which are isomorphic to a matrix group. We need
some facts about groups of matrices, which we shall compile in
this section. For a more general treatment, see Chevalley (12).

Let $\mathfrak{G}$ be a group of $n \times n$ matrices of dimension r, that is,
each matrix $u \in \mathfrak{G}$ depends on r independent parameters a_1,
$a_2, \cdots, a_r$; more precisely, each matrix $u \in \mathfrak{G}$ is determined by a
point $a = (a_1, a_2, \cdots, a_r)$ of a differentiable manifold of dimension
r, which we shall denote by the same letter $\mathfrak{G}$; $a_1, a_2, \cdots, a_r$ are then
the coordinates of a in a suitable local coordinate system.

Let $e \in \mathfrak{G}$ be the unit matrix and u^{-1} the inverse of $u \in \mathfrak{G}$. If
du denotes the differential of the matrix u, the equation

$$(2.1) \qquad\qquad u^{-1}(u + du) = e + \omega$$

defines a matrix $\omega = u^{-1} du$ of linear (pfaffian) differential forms
which is called the matrix of Maurer-Cartan of $\mathfrak{G}$. The elements
ω_{ij} of ω have the form $\omega_{ij} = \alpha_{ij1} da_1 + \cdots + \alpha_{ijr} da_r$, where the
coefficients α_{ijk} are analytic functions of $a_1, a_2, \cdots, a_r$. From these
n^2 pfaffian forms ω_{ij} there are r linearly independent (base of the
vector space dual of the tangent space of $\mathfrak{G}$) which we shall denote by $\omega_1, \omega_2, \cdots, \omega_r$; they are called the forms of Maurer-Cartan

of $\mathfrak{G}$ and are defined up to a linear combination with constant coefficients.

The fundamental property of the matrix ω is that of being left invariant under $\mathfrak{G}$. For if $u' = su$ (s is a fixed element of $\mathfrak{G}$), we have $du' = s\,du$, and therefore $\omega' = u'^{-1}\,du' = u^{-1}\,s^{-1}\,s\,du = u^{-1}\,du = \omega$.

As a consequence, the r forms of Maurer-Cartan are also left invariant under $\mathfrak{G}$, and this fact characterizes these forms up to a linear combination with constant coefficients. For a proof, we observe that since the forms of Maurer-Cartan $\omega_1, \cdots, \omega_r$ are independent, each pfaffian form Ω may be written $\Omega(a, da) = \Sigma_1^r A_i(a)\omega_i$. If Ω is left invariant under $\mathfrak{G}$, we have

$$\Omega' = \Sigma_1^r A_i(a')\omega_i' = \Sigma_1^r A_i(a)\omega_i$$

and since $\omega_i' = \omega_i$, we have

$$\Sigma_1^r (A_i(a') - A_i(a))\omega_i = 0.$$

Because of the independence of ω_i, it follows that $A_i(a') = A_i(a)$, which implies $A_i =$ constant. (Since we are interested only in the left invariance, we shall hereafter speak simply of invariance, understanding that it means left invariance.)

Notice that by exterior differentiation of $\omega = u^{-1}\,du$, taking into account that $du^{-1} = -u^{-1}\,du\,u^{-1}$, we get

$$(2.2) \qquad d\omega = -u^{-1}\,du\,u^{-1} \wedge du = -\omega \wedge \omega.$$

This matric equation includes the expression of the exterior differentials $d\omega_i$ of the forms of Maurer-Cartan as linear combinations with constant coefficients of the products $\omega_j \wedge \omega_k$; these expressions are called the equations of structure of Maurer-Cartan for the group $\mathfrak{G}$.

2.2. *Density and measure in homogeneous spaces.* Let $\mathfrak{H}$ be a subgroup of $\mathfrak{G}$ of dimension $r - h$. Suppose that $\mathfrak{H}$ itself is a Lie group isomorphic to a matrix group. We want to find the conditions for the existence of a density (that is, an element of volume) in the homogeneous space $\mathfrak{G}/\mathfrak{H}$ (= set of left cosets $s\mathfrak{H}$, $s \in \mathfrak{G}$) invariant under $\mathfrak{G}$. For this purpose, we notice that the submanifold $\mathfrak{H}$ of the differentiable manifold $\mathfrak{G}$ and its left cosets $s\mathfrak{H}$ $(s \in \mathfrak{G})$ are the integral manifolds of a pfaffian system.

$$(2.3) \qquad \omega_1 = 0, \qquad \omega_2 = 0, \qquad \cdots, \qquad \omega_h = 0.$$

Because $\mathfrak{H}$ and its left cosets as a whole are invariant under $\mathfrak{G}$, the left side members of (2.3) will be linear combinations with constant coefficients of the forms of Maurer-Cartan of $\mathfrak{G}$, and, because these forms are defined up to a linear combination with constant coefficients, we may assume that they are the h first forms of Maurer-Cartan of $\mathfrak{G}$.

Because ω_i is invariant under $\mathfrak{G}$, the differential form

$$(2.4) \qquad \Omega_h = \omega_1 \wedge \omega_2 \wedge \cdots \wedge \omega_h$$

will be also invariant under $\mathfrak{G}$. However, Ω_h is not always a density for $\mathfrak{G}/\mathfrak{H}$ because its value can change when the points $a \in \mathfrak{G}$ displace on the manifolds $s\mathfrak{H}$. We shall now prove the following theorem.

THEOREM: *A necessary and sufficient condition for Ω_h to be a density for $\mathfrak{G}/\mathfrak{H}$ is that its exterior differential vanish, that is,*

$$(2.5) \qquad d\Omega_h = 0.$$

Proof: To prove this theorem, we observe that the submanifold $\mathfrak{H}$ and its left cosets fill up the manifold $\mathfrak{G}$ in such a way that for each point of $\mathfrak{G}$ passes one and only one submanifold. Thus, the system (2.3) is completely integrable and it is consequently equivalent to a system of the form

$$(2.6) \qquad d\xi_1 = 0, \qquad d\xi_2 = 0, \qquad \cdots, \qquad d\xi_h = 0,$$

where $\xi_i = \xi_i(a_1, a_2, \cdots, a_r)$ are functions of a_i such that the manifolds $s\mathfrak{H}$ are represented by $\xi_i = $ constant $(i = 1, 2, \cdots, h)$. We can make in $\mathfrak{G}$ the change of local coordinates $(a_1, a_2, \cdots, a_r) \rightarrow (\xi_1, \xi_2, \cdots, \xi_h, x_{h+1}, \cdots, x_r)$. Since the systems (2.3) and (2.6) are equivalent, we have

$$(2.7) \qquad \Omega_h = A(\xi, x)\, d\xi_1 \wedge d\xi_2 \wedge \cdots \wedge d\xi_h,$$

where $A(\xi, x)$ denotes a function of $\xi_1, \cdots, \xi_h, x_{h+1}, \cdots, x_r$. When the point $a(\xi_1, \xi_2, \cdots, \xi_h, x_{h+1}, \cdots, x_r)$ varies on $s\mathfrak{H}$, the coordinates ξ_i are constant, and, therefore,

(2.8) $$\delta\Omega_h = \sum_{j=h+1}^{r} \frac{\partial A}{\partial x_j} \, dx_j \wedge d\xi_1 \wedge \cdots \wedge d\xi_h.$$

On the other side, by exterior differentiation of (2.7), we get

$$d\Omega_h = \sum_{j=1}^{h} \frac{\partial A}{\partial \xi_j} \, d\xi_j \wedge d\xi_1 \wedge \cdots \wedge d\xi_h$$

$$+ \sum_{j=h+1}^{r} \frac{\partial A}{\partial x_j} \, dx_j \wedge d\xi_1 \wedge \cdots \wedge d\xi_h = \delta\Omega_h,$$

because the first sum vanishes. Consequently, so that $\delta\Omega_h = 0$— that is, for Ω_h to be invariant by displacements on the manifolds $s\mathfrak{H}$, it is necessary and sufficient that $d\Omega_h = 0$. This proves the theorem.

If $\mathfrak{H}$ reduces to the identity, then $\mathfrak{G}/\mathfrak{H} = \mathfrak{G}$ and $\Omega_r = \omega_1 \wedge \omega_2 \wedge \cdots \wedge \omega_r$ gives the invariant density ($=$ element of volume) of $\mathfrak{G}$, which in integral geometry takes the name of kinematic density of $\mathfrak{G}$. The integral of Ω_r gives an invariant measure for $\mathfrak{G}$ (Haar's measure) which is unique up to a constant factor.

2.3. *The examples of the introduction.* To exemplify these general results, we shall consider the examples appearing in the introduction.

The group of motions $\mathfrak{G} = \mathfrak{M}$ in E_2 can be represented by the group of 3-dimensional matrices,

(2.9) $$u = \begin{pmatrix} \cos\varphi & -\sin\varphi & a \\ \sin\varphi & \cos\varphi & b \\ 0 & 0 & 1 \end{pmatrix}$$

with the parameters $a_1 = a$, $a_2 = b$, $a_3 = \varphi$. We have

$$u^{-1} = \begin{pmatrix} \cos\varphi & \sin\varphi & -b\sin\varphi - a\cos\varphi \\ -\sin\varphi & \cos\varphi & -b\cos\varphi + a\sin\varphi \\ 0 & 0 & 1 \end{pmatrix}$$

$$du = \begin{pmatrix} -\sin\varphi \, d\varphi & -\cos\varphi \, d\varphi & da \\ \cos\varphi \, d\varphi & -\sin\varphi \, d\varphi & db \\ 0 & 0 & 0 \end{pmatrix}$$

and, therefore,

$$\omega = u^{-1}\, du = \begin{pmatrix} 0 & -d\varphi & \cos\varphi\, da + \sin\varphi\, db \\ d\varphi & 0 & -\sin\varphi\, da + \cos\varphi\, db \\ 0 & 0 & 0 \end{pmatrix}$$

The forms of Maurer-Cartan are

(2.10)
$$\omega_1 = \cos\varphi\, da + \sin\varphi\, db, \quad \omega_2 = -\sin\varphi\, da + \cos\varphi\, db, \quad \omega_3 = d\varphi,$$

and the equations of structure

$$d\omega = -\omega \wedge \omega = -\begin{pmatrix} 0 & 0 & -\omega_3 \wedge \omega_2 \\ 0 & 0 & \omega_3 \wedge \omega_1 \\ 0 & 0 & 0 \end{pmatrix}.$$

That is,

(2.11) $d\omega_1 = -\omega_2 \wedge \omega_3, \qquad d\omega_2 = -\omega_3 \wedge \omega_1, \qquad d\omega_3 = 0.$

The kinematic density of $\mathfrak{M}$ is

$$dK = \omega_1 \wedge \omega_2 \wedge \omega_3 = da \wedge db \wedge d\varphi,$$

which, up to the notation, coincides with (1.11).

Let $\mathfrak{H}_1$ be the subgroup of $\mathfrak{M}$ consisting of all motions which leave the line $G(p, \theta)$ invariant (equation of $G: x \cos\theta + y \sin\theta - p = 0$). There is a bijective mapping between the lines G of E_2 and the points of the space $\mathfrak{M}/\mathfrak{H}_1$. As density for lines, we take the density of $\mathfrak{M}/\mathfrak{H}_1$.

By the change of coordinates $(a, b, \varphi) \to (p, \theta, t)$ in $\mathfrak{M}$, given by the equations,

$$a = p \cos\theta + t \sin\theta, \qquad b = p \sin\theta - t \cos\theta, \qquad \varphi = \theta - \frac{\pi}{2}$$

$$p = a \cos\theta + b \sin\theta, \qquad t = a \sin\theta - b \cos\theta, \qquad \theta = \varphi + \frac{\pi}{2},$$

the points of $\mathfrak{M}/\mathfrak{H}_1$ are $p = $ constant, $\theta = $ constant. The system (2.6) is $dp = 0$, $d\theta = 0$, and the system (2.3) is

$$dp = \cos\theta\, da + \sin\theta\, db = -\sin\varphi\, da + \cos\varphi\, db = \omega_2 = 0,$$

$$d\theta = d\varphi = \omega_3 = 0.$$

Therefore, the density for lines takes the form

(2.12) $dG = \omega_2 \wedge \omega_3 = -\sin\varphi\, da \wedge d\varphi + \cos\varphi\, db \wedge d\varphi$

which is equivalent to

$$(2.13) \qquad dG = dp \wedge d\theta,$$

as stated in (1.7).

If $\mathfrak{H}_0$ is the subgroup of $\mathfrak{M}$ consisting of all motions which leave the point $P(a, b)$ invariant, there is a bijective mapping between the points (a, b) of E_2 and the points of the homogeneous space $\mathfrak{M}/\mathfrak{H}_0$. The system (2.6) is now $da = 0$, $db = 0$, and (2.3) gives $\omega_1 = 0$, $\omega_2 = 0$. The density (2.4) for points results in

$$(2.14) \qquad dP = \omega_1 \wedge \omega_2 = da \wedge db,$$

which coincides with (1.4). In both cases (2.13) and (2.14), the condition (2.5) is obviously satisfied.

To give an example in which the homogeneous space $\mathfrak{G}/\mathfrak{H}$ has not an invariant density, let us consider the 4-dimensional group $\mathfrak{G}$ of matrices of the form

$$u = \begin{pmatrix} a_1 & 0 & a_2 \\ 0 & a_3 & a_4 \\ 0 & 0 & 1 \end{pmatrix}, \quad a_1 a_3 \neq 0,$$

and the 2-dimensional subgroup $\mathfrak{H}$ of matrices of the form

$$u_1 = \begin{pmatrix} a_1 & 0 & 0 \\ 0 & a_3 & 0 \\ 0 & 0 & 1 \end{pmatrix}, \quad a_1 a_3 \neq 0.$$

To obtain the forms of Maurer-Cartan of $\mathfrak{G}$, we have

$$\omega = u^{-1} \, du = \begin{pmatrix} a_1^{-1} & 0 & -a_1^{-1}a_2 \\ 0 & a_3^{-1} & -a_3^{-1}a_4 \\ 0 & 0 & 1 \end{pmatrix} \begin{pmatrix} da_1 & 0 & da_2 \\ 0 & da_3 & da_4 \\ 0 & 0 & 0 \end{pmatrix}$$

$$= \begin{pmatrix} \omega_1 & 0 & \omega_2 \\ 0 & \omega_3 & \omega_4 \\ 0 & 0 & 0 \end{pmatrix},$$

where

$$\omega_1 = a_1^{-1} \, da_1, \qquad \omega_2 = a_1^{-1} \, da_2, \qquad \omega_3 = a_3^{-1} \, da_3, \qquad \omega_4 = a_3^{-1} \, da_4.$$

The subgroup $\mathfrak{H}$ is characterized by $a_2 = 0$, $a_4 = 0$, and, therefore, the system (2.3) is now $\omega_2 = 0$, $\omega_4 = 0$. The differential form

$\Omega_2 = \omega_2 \wedge \omega_4$ is not a density, because $d\Omega_2 = -\omega_1 \wedge \omega_2 \wedge \omega_4 - \omega_3 \wedge \omega_2 \wedge \omega_4 \neq 0$.

3. INTEGRAL GEOMETRY IN THE THREE-DIMENSIONAL EUCLIDEAN SPACE

3.1. *The group of motions in E_3.* We shall consider in detail the integral geometry of the 3-dimensional euclidean space. The base space is E_3 and the group $\mathfrak{G}$ is the group of motions $\mathfrak{M}$ in it.

Let x represent the one-column matrix formed by the orthogonal coordinates x_1, x_2, x_3 of a point P. The matrix equation of a motion $x \to x'$ is

$$(3.1) \qquad\qquad x' = Ax + B,$$

where

$$(3.2) \qquad A = \begin{pmatrix} a_{11} & a_{12} & a_{13} \\ a_{21} & a_{22} & a_{23} \\ a_{31} & a_{32} & a_{33} \end{pmatrix}, \quad B = \begin{pmatrix} b_1 \\ b_2 \\ b_3 \end{pmatrix},$$

and A satisfies the conditions of orthogonality

$$(3.3) \qquad A^t = A^{-1} \quad (A^t = \text{transposed of } A).$$

The condition (3.3) reduces to 3 the number of independent parameters a_{ij} which, with b_1, b_2, and b_3, are the 6 parameters on which $\mathfrak{M}$ depends.

The group $\mathfrak{M}$ can be represented by the 4×4 matrices,

$$(3.4) \qquad u = \begin{pmatrix} A & \vdots & B \\ \cdots\cdots\cdots\cdots \\ 0 & \vdots & 1 \end{pmatrix}$$

with the ordinary rules,

$$u_2 u_1 = \begin{pmatrix} A_2 & A_1 & \vdots & A_2 B_1 + B_2 \\ \cdots\cdots\cdots\cdots\cdots \\ 0 & \vdots & 1 \end{pmatrix}, \qquad u^{-1} = \begin{pmatrix} A^{-1} & \vdots & -A^{-1}B \\ \cdots\cdots\cdots\cdots \\ 0 & \vdots & 1 \end{pmatrix}.$$

The matrix of Maurer-Cartan is

$$\omega = u^{-1}\, du = \begin{pmatrix} A^{-1}\, dA & \vdots & A^{-1}\, dB \\ \cdots\cdots\cdots\cdots\cdots \\ 0 & \vdots & 0 \end{pmatrix}.$$

If we introduce the two matrices

(3.5) $\omega_A = A^{-1}\, dA, \qquad \omega_B = A^{-1}\, dB$

of order 3×3 and 3×1, respectively, the equations of structure can be written

(3.6) $d\omega_A = -\omega_A \wedge \omega_A, \qquad d\omega_B = -\omega_A \wedge \omega_B.$

Since $\mathfrak{M}$ is a 6-parameter group, we must have 6 pfaffian forms of Maurer-Cartan. Effectively, from (3.3) and (3.5) we deduce $\omega_A = A^t\, dA = -dA^t\, A = \omega_A^t$, and the 6 forms are the elements of the matrices,

$$\omega_A = \begin{pmatrix} 0 & \omega_{12} & \omega_{13} \\ -\omega_{12} & 0 & \omega_{23} \\ -\omega_{13} & -\omega_{23} & \omega_{33} \end{pmatrix}, \qquad \omega_B = \begin{pmatrix} \omega_1 \\ \omega_2 \\ \omega_3 \end{pmatrix},$$

which, explicitly, give

(3.7) $\omega_{ih} = -\omega_{hi} = \sum\limits_{j=1}^{3} a_{ji}\, da_{jh}, \qquad \omega_i = \sum\limits_{j=1}^{3} a_{ji}\, db_j.$

It is useful to give a more geometrical approach to the pfaffian forms ω_{ih} and ω_i. Let us consider in E_3 a fixed frame $(Q_0; e_1^0, e_2^0, e_3^0)$ composed of a point Q_0 and three orthogonal unit vectors e_i^0, and a moving frame $(Q; e_1, e_2, e_3)$ which results from the fixed frame by the motion u represented by (3.1). If we introduce the matrices

(3.8) $e^0 = (e_1^0, e_2^0, e_3^0), \qquad e = (e_1, e_2, e_3)$

we can write

(3.9) $Q = e^0 B, \qquad e = e^0 A,$

and, therefore,

(3.10)
$dQ = e^0\, dB = e\, A^{-1}\, dB = e\omega_B,$

$de = e^0\, dA = e\, A^{-1}\, dA = e\omega_A,$

which may be written

$$(3.11) \qquad dQ = \sum_{j=1}^{3} \omega_j e_j, \quad de_i = \sum_{j=1}^{3} \omega_{ji} e_j.$$

These formulas are useful for the computation of densities, as we shall see in the next section. Because of the orthogonality of the unit vectors e_i, we have $e_i e_j = \delta_{ij}$, and from (3.11) we deduce

$$(3.12) \qquad \omega_j = e_j \, dQ, \quad \omega_{ji} = e_j \, de_i,$$

which are the vectorial form of the equations in (3.7).

3.2. *The area element of the unit sphere.* We need to remember two expressions for the element of area of the unit sphere. Let ν be the unit vector with the components

$$(3.13) \quad \nu_1 = \sin \theta \cos \varphi, \quad \nu_2 = \sin \theta \sin \varphi, \quad \nu_3 = \cos \theta$$

where θ, φ are the ordinary spherical coordinates corresponding to the endpoint of ν. The area element at this endpoint is known to be

$$(3.14) \qquad d\sigma = (\nu \, \nu_\theta \, \nu_\varphi) \, d\theta \wedge d\varphi = \sin \theta \, d\theta \wedge d\varphi$$

where $(\nu \, \nu_\theta \, \nu_\varphi)$ denotes the scalar triple product of the vectors ν, ν_θ, and ν_φ (subscripts denote partial derivation). Taking (3.13) into account, we have also

$$(3.15) \qquad d\sigma = \frac{d\nu_2 \wedge d\nu_3}{\nu_1} = \frac{d\nu_3 \wedge d\nu_1}{\nu_2} = \frac{d\nu_1 \wedge d\nu_2}{\nu_3},$$

and since $\nu_1^2 + \nu_2^2 + \nu_3^2 = 1$, we deduce

$$d\sigma = \nu_1 \, d\nu_2 \wedge d\nu_3 + \nu_2 \, d\nu_3 \wedge d\nu_1 + \nu_3 \, d\nu_1 \wedge d\nu_2.$$

On the other hand, if e_1, e_2, and e_3 are the 3 orthogonal unit vectors of a moving frame, we have

$$(3.16) \quad \begin{aligned} e_1 \, de_3 \wedge e_2 \, de_3 &= e_1(e_{3\theta} \, d\theta + e_{3\varphi} \, d\varphi) \wedge e_2(e_{3\theta} \, d\theta + e_{3\varphi} \, d\varphi) \\ &= (e_1 e_{3\theta} \cdot e_2 e_{3\varphi} - e_1 e_{3\varphi} \cdot e_2 e_{3\theta}) \, d\theta \wedge d\varphi \\ &= (e_1 \wedge e_2) \cdot (e_{3\theta} \wedge e_{3\varphi}) \, d\theta \wedge d\varphi \\ &= (e_3 e_{3\theta} e_{3\varphi}) \, d\theta \wedge d\varphi = d\sigma \end{aligned}$$

where $d\sigma$ denotes the area element of the unit sphere corresponding to the endpoint of e_3. From (3.12) and (3.16), we get

(3.17) $$d\sigma = \omega_{13} \wedge \omega_{23}.$$

We have now at our disposal all elements necessary to find the densities for points, lines and planes of E_3 invariant under $\mathfrak{M}$.

3.3. *Density for points.* Let $\mathfrak{H}_0$ be the set of motions which leave the point $Q(b_1, b_2, b_3)$ invariant; clearly it is a subgroup of $\mathfrak{M}$. According to (3.11), to keep Q fixed we must have

$$\omega_1 = 0, \qquad \omega_2 = 0, \qquad \omega_3 = 0,$$

which is the system (2.3), and, according to (2.4), the density for points will be $\omega_1 \wedge \omega_2 \wedge \omega_3 = db_1 \wedge db_2 \wedge db_3$ [applying (3.7) and taking into account the determinant $|a_{ij}| = 1$, because the matrix $A = (a_{ij})$ is orthogonal]. In general, for the point $P(x, y, z)$, we shall have

(3.18) $$dP = dx \wedge dy \wedge dz.$$

The condition (2.5) is obviously satisfied.

3.4. *Density for planes.* Let $\mathfrak{H}_2$ be the set of motions which leave the plane $E(e_1, e_2)$ invariant; clearly it is a subgroup of $\mathfrak{M}$.

By the motions of $\mathfrak{H}_2$ the unit vector e_3 remains fixed and the point Q can only move on the plane e_1, e_2; therefore, according to (3.11), the pfaffian system which characterizes the planes is

$$\omega_3 = 0, \qquad \omega_{13} = 0, \qquad \omega_{23} = 0,$$

and the density for planes results:

(3.19) $$dE = \omega_3 \wedge \omega_{13} \wedge \omega_{23}.$$

If θ, φ are the spherical coordinates of the endpoint of e_3, (3.14) and (3.17) give

(3.20) $$\omega_{13} \wedge \omega_{23} = d\sigma = \sin \theta \, d\theta \wedge d\varphi.$$

If p is the distance from the origin Q_0 of the fixed frame to the plane E, and $a_{13} = \sin \theta \cos \varphi$, $a_{23} = \sin \theta \sin \varphi$, $a_{33} = \cos \theta$ are the components of e_3 (normal to E), we have $p = a_{13}b_1 + a_{23}b_2 + a_{33}b_3$, and, according to (3.7),

(3.21) $$\omega_3 = \sum_{j=1}^{3} a_{j3} \, db_j = dp + R \, d\theta + S \, d\varphi.$$

Here, R, S are functions of θ, φ, b_i, the explicit form of which has no interest for us. From (3.19) and (3.20) we get

$$(3.22) \qquad dE = \sin\theta\, dp \wedge d\theta \wedge d\varphi = dp \wedge d\sigma.$$

The condition (2.5) is obviously satisfied, and hence we have: If a plane E is determined by its normal e_3 and its distance p to a fixed origin, the density is given by (3.22), where $d\sigma$ denotes the area element of the unit sphere corresponding to the endpoint of the unit vector e_3.

As an exercise, prove that if the plane is given by the equation $ux + vy + wz + 1 = 0$, its density takes the form

$$dE = \frac{du \wedge dv \wedge dw}{(u^2 + v^2 + w^2)^2}.$$

Example

Let S be a fixed segment of length l. To compute the measure of the set of planes E which intersect S, we take S on the e_3^0-axis and the middle point of S as the origin of coordinates. Then we have

$$(3.23) \quad m(E; E \cap S \neq 0) = \int\limits_{E \cap S \neq 0} dE$$

$$= \frac{l}{2} \int_0^{2\pi} d\varphi \int_0^\pi |\cos\theta|\, \sin\theta\, d\theta = \pi l.$$

If Γ is a polygonal line of length L, writing (3.23) for all sides of Γ and adding, we obtain

$$(3.24) \qquad\qquad \int n\, dE = \pi L,$$

where n denotes the number of intersection points of E with Γ. By a limit process it is not difficult to prove that (3.24) holds for any rectifiable curve. The integral in (3.24) is extended over all planes of E_3, n being 0 for the planes which do not intersect Γ.

3.5. *Density for straight lines.* Let $\mathfrak{H}_1$ be the set of motions leaving the line G which contains the unit vector e_3 invariant; clearly $\mathfrak{H}_1$ is a subgroup of $\mathfrak{M}$.

By a motion of $\mathfrak{H}_1$, the point Q can only move in the direction of e_3, and, therefore, (3.11) gives $\omega_1 = 0$, $\omega_2 = 0$. Moreover, because e_3 is fixed, from (3.11) we deduce $\omega_{13} = 0$, $\omega_{23} = 0$. The pfaffian system (2.3) for the lines of E_3 becomes

$$(3.25) \qquad \omega_1 = 0, \qquad \omega_2 = 0, \qquad \omega_{13} = 0, \qquad \omega_{23} = 0,$$

and the density for lines is

$$(3.26) \qquad dG = \omega_1 \wedge \omega_2 \wedge \omega_{13} \wedge \omega_{23}.$$

According to (3.12), $\omega_1 \wedge \omega_2$ equals the area element of the plane (e_1, e_2) at the point Q, and we have seen that $\omega_{13} \wedge \omega_{23}$ is the area element of the unit sphere corresponding to the endpoint of e_3, that is, to the direction of G. If G is determined by its direction e_3 and its intersection point (x, y) with a fixed plane, denoting by ψ the angle between e_3 and the normal to the fixed plane, we have $\omega_1 \wedge \omega_2 = |\cos \psi| \, dx \wedge dy$, and we can write (3.26) in the form

$$(3.27) \qquad dG = |\cos \psi| \, dx \wedge dy \wedge d\sigma.$$

From (3.26) and (3.6) it is easy to show that the condition (2.5) is satisfied.

As an exercise, prove that if G is given by the equations $x = az + p$, $y = bz + q$, then its density is

$$dG = \frac{da \wedge db \wedge dp \wedge dq}{(1 + a^2 + b^2)^2}.$$

Example

Let Σ be a fixed surface of class C^1 (= with a continuous tangent plane). If P denotes a point of the intersection $G \cap \Sigma$ and df denotes the area element of Σ at P, the density for lines can be written $dG = |\cos \psi| \, df \wedge d\sigma$, where ψ denotes the angle between G and the normal to Σ at P. Fixed P, the integral of $|\cos \psi| \, d\sigma$ extended over all the lines which pass through P, gives the projection of one-half the unit sphere upon a diametral plane—that is, π. The integration of df over the whole Σ gives the area F of Σ. Therefore, taking into account that each line has been counted as many times n as it has intersection points with Σ, we get

$$(3.28) \qquad \int n \, dG = \pi F,$$

where the integral is extended over all lines of E_3, n being 0 for the lines which do not intersect Σ.

3.6. *Kinematic density.* The kinematic density is

$$(3.29) \qquad dK = \omega_1 \wedge \omega_2 \wedge \omega_3 \wedge \omega_{12} \wedge \omega_{13} \wedge \omega_{23}.$$

To give a geometrical interpretation to $\omega_{12} = e_1 \, de_2$, we observe that if we take on the plane e_1, e_2 two fixed orthogonal unit vectors e_1^*, e_2^* and call α the angle between e_1 and e_1^*, we can write $e_1 = \cos \alpha \, e_1^* + \sin \alpha \, e_2^*$, $e_2 = -\sin \alpha \, e_1^* + \cos \alpha \, e_2^*$; therefore, $e_1 \, de_2 = -d\alpha$. That is, ω_{12} means an elementary rotation about the e_3-axis. Consequently, according to (3.17) and (3.29), if a motion is determined by the position of the moving frame $(Q; e_1, e_2, e_3)$, the kinematic density has the form

$$(3.30) \qquad dK = dP \wedge d\sigma \wedge d\alpha,$$

where dP is the volume element of E_3 at the origin Q of the moving frame, $d\sigma$ is the area element of the unit sphere corresponding to the endpoint of e_3, and $d\alpha$ is the element of rotation about e_3. We remember that we always consider the densities in absolute value; thus, there is no question of sign.

Let us do an application of (3.30). Let Γ be a fixed curve with continuous tangent at every point and finite length L and let Σ be a moving surface of class C^1 and finite area F. Let Q be a point of $\Gamma \cap \Sigma$ and let e_3 be the normal to Σ at Q. If θ denotes the angle between e_3 and the tangent to Γ at Q (which we may take as the e_3^0-axis of the fixed frame) and df denotes the area element of Σ at Q, we have $dP = |\cos \theta| \, df \wedge ds$ (s = arc length of Γ). Putting this value in (3.30) and integrating over all the positions of Σ in which it has common point with Γ, because each position of Σ will be counted as many times n as intersection points have Σ and Γ, we get

$$(3.31) \qquad \int n \, dK = 4\pi^2 FL.$$

Notice that the same formula holds if we suppose Σ fixed and Γ moving with density dK.

If Σ is the unit sphere, we can take the origin of the moving frame at the center of Σ; then we have $\int n\, dK = 8\pi^2 \int n\, dP$, and (3.31) gives

$$(3.32) \qquad\qquad \int n\, dP = 2\pi L,$$

which is valid for any rectifiable curve (51).

3.7. *A differential formula.* In Section 5 we will need an important auxiliary formula which derives from (3.30). Let Σ_0 be a fixed surface of class C^1. At each point Q of Σ_0 we consider an orthogonal frame $(Q; e_1^0, e_2^0, e_3^0)$ with origin at Q and with e_3^0 normal to Σ_0. If the displacement vector on Σ_0 at Q is $\omega_1 e_1^0 + \omega_2 e_2^0$, the area element is $df = \omega_1 \wedge \omega_2$. To the unit vector e^0 tangent to Σ_0 at Q which forms with e_1^0 the angle τ_0, is attached the differential form $dL_0 = \omega_1 \wedge \omega_2 \wedge d\tau_0$ called the density for line elements $(Q; e^0)$ on Σ_0, and the pfaffian form $d\theta = \cos \tau_0\, \omega_1 + \sin \tau_0\, \omega_2$ called the element of length corresponding to the direction e^0.

Now let Σ_1 be a moving surface of class C^1, and assume that the intersection $\Sigma_0 \cap \Sigma_1$ is a rectifiable curve Γ. Let Q be a point of Γ and $(Q; e_1, e_2, e_3)$ be an orthogonal frame with e_3 perpendicular to Σ_1. Let ds be the length element of Γ at Q and ds_0, ds_1 those normal to Γ on Σ_0 and Σ_1, respectively. Let θ be the angle between the normals e_3^0, e_3. If df_0, df_1 are the elements of area of Σ_0, Σ_1 at Q and dP denotes the element of volume of E_3 at Q, we have $dP = \sin\theta\, df_0 \wedge ds_1$ and $df_1 = ds \wedge ds_1$. The element of area of the unit sphere at the endpoint of e_3 may be written $d\sigma = \sin\theta\, d\theta \wedge d\tau_0$. Putting now $d\tau_1 = d\alpha$ to unify the notation of (3.30), from this equation and the preceding relation, we deduce immediately (up to the sign)

$$(3.33) \qquad ds \wedge dK = \sin^2\theta\, df_0 \wedge d\tau_0 \wedge df_1 \wedge d\tau_1 \wedge d\theta$$
$$= \sin^2\theta\, dL_0 \wedge dL_1 \wedge d\theta,$$

which is the differential formula we want.

An immediate consequence is obtained by integrating both sides over all positions of the moving surface Σ_1. We get

$$(3.34) \qquad\qquad \int L\, dK = 2\pi^3 F_0 F_1,$$

where L denotes the length of the curve $\Sigma_0 \cap \Sigma_1$, and F_0, F_1 are the areas of Σ_0, Σ_1, respectively.

If Σ_1 is the unit sphere and we take the origin of the moving frame at the center of Σ_1, we have

$$\int L \, dK = 8\pi^2 \int L \, dP_1$$

and (3.34) gives

$$(3.35) \qquad \int L \, dP = \pi^2 F_0.$$

3.8. *A definition of area.* Let now Σ_1, Σ_2 be two moving unit spheres and Σ_0 a fixed surface. Let N be the number of points of the intersection $\Sigma_0 \cap \Sigma_1 \cap \Sigma_2$. If dP_i denotes the volume element at the center of Σ_i $(i = 1, 2)$, we get from (3.32) and (3.35)

$$\int N \, dP_1 \, dP_2 = 2\pi \int L \, dP_1 = 2\pi^3 F_0.$$

Conversely, this result conduces to define the area of a continuum of points by the formula

$$F_0 = \frac{1}{2\pi^3} \int N \, dP_1 \, dP_2,$$

provided the integral of the right-hand side exists [see (52)]. Applications of the integral geometry to the definition of area for k-dimensional surfaces have been made by Federer (17–19) and Hadwiger (23) and (25). See also Nöbeling (45) and (46).

3.9. *Planes through a fixed point.* Let us now consider the set of planes E_0 which pass through a fixed point O. The density for sets of E_0 invariant under the group $\mathfrak{M}_0$ of the rotations about 0, is clearly $dE_0 = d\sigma$, where $d\sigma$ denotes the area element of the unit sphere corresponding to the direction perpendicular to E_0. In fact, this differential form is invariant under $\mathfrak{M}_0$, and, because of the transitivity of the planes E_0 with respect to $\mathfrak{M}_0$, it is unique up to a constant factor. The planes E_0 are considered non-oriented; therefore, the measure of all the planes through O will be

$$(3.36) \qquad \int dE_0 = \int_{\frac{1}{2}Z} d\sigma = 2\pi,$$

where $\frac{1}{2}Z$ denotes the half of the unit sphere.

Let S be a fixed arc of great circle on the unit sphere of center O of length α. The measure of the set of planes E_0 which intersect S (= measure of the set of great circles which intersect S) will be the area of the lune bounded by the great circles the poles of which are the endpoints of S—that is, $m(E_0; S \cap E_0 \neq 0) = 2\alpha$. If instead of S we have a spherical polygonal line Γ the sides of which have the lengths α_i, we have, writing the last formula for each side and adding,

$$(3.37) \qquad \int n \, dE_0 = 2L,$$

where L denotes the total length of Γ. The integration is extended over all (non-oriented) planes through O—that is, according to $dE_0 - d\sigma$, over half the unit sphere. By a limit process we can prove that (3.37) holds for any rectifiable spherical curve of the unit sphere.

Following Fenchel (20), we want to apply (3.37). Let K be a closed space curve of class C^2 without multiple points and let Γ be the spherical indicatrix of it (= the curve $T = T(s)$, where T is the tangent unit vector to K). The arc length element of Γ is $ds_t = |\varkappa| \, ds$, where $\varkappa$ denotes the curvature and s the length of K. Consequently, (3.37) yields

$$(3.38) \qquad \int n \, dE_0 = 2 \int_K |\varkappa| \, ds.$$

Every closed space curve K has at least 2 tangents which are parallel to an arbitrary plane. This means that every plane E_0 intersects Γ in at least 2 points. Hence, $n \geqq 2$, and (3.36) and (3.38) give

$$(3.39) \qquad \int_K |\varkappa| \, ds \geqq 2\pi,$$

a classical inequality of Fenchel.

If K is knotted, it is easy to see that it has at least 4 tangents parallel to an arbitrary plane. Hence, $n \geqq 4$, and (3.36) and (3.38) give the following inequality of Fáry (for knotted curves) (16),

$$(3.40) \qquad \int_K |\varkappa| \, ds \geqq 4\pi.$$

These results have been generalized to closed varieties in E_n by Chern and Lashof (11).

4. APPLICATIONS TO CONVEX BODIES

The integral geometry is closely related to the theory of convex bodies. We compile in this section some simple facts on this theory from many sources—for example, Bonnesen and Fenchel (4), Busemann (5), Hadwiger (24), and Vincensini (72).

Let k be a plane convex set of area f placed in E_3. Let f_σ be the area of the orthogonal projection of k on a plane perpendicular to the direction σ, and let θ be the angle between σ and the normal to the plane which contains k; we have $f_\sigma = |\cos \theta| f$. If $d\sigma$ denotes the area element of the unit sphere Z corresponding to the direction σ, we have

$$(4.1) \qquad \int_Z f_\sigma \, d\sigma = f \int_0^{2\pi} d\varphi \int_0^\pi |\cos \theta| \sin \theta \, d\theta = 2\pi f,$$

and, therefore,

$$(4.2) \qquad f = \frac{1}{2\pi} \int_Z f_\sigma \, d\sigma.$$

Now let K be a convex body of E_3; we shall denote by ∂K the convex surface bounding K. Let F be the area of ∂K and F_σ the area of the orthogonal projection of K on a plane perpendicular to the direction σ. Applying (4.2) to each element of area of ∂K and integrating over all ∂K, we get

$$(4.3) \qquad F = \frac{1}{\pi} \int_Z F_\sigma \, d\sigma,$$

known as Cauchy's formula for the area of a convex body.

Let O be an interior point of K and $p = p(\sigma) = p(\theta, \varphi)$ be the supporting function of K with respect to O (= distance from O to the supporting plane perpendicular to the direction σ of spherical coordinates θ, φ). The convex body K_h parallel to K at distance h has the supporting function $p_h = p(\sigma) + h$, and if R_1, R_2 are the principal radii of curvature of ∂K, those of ∂K_h at corresponding points are $R_1 + h$ and $R_2 + h$. Between the area element df of

∂K and the area element $d\sigma$ of its spherical image, there is the relation $df/d\sigma = R_1 R_2$, and consequently, we have

$$(4.4) \qquad F = \int_Z R_1 R_2 \, d\sigma.$$

Applying this formula to ∂K_h, we get

$$(4.5) \quad F_h = \int_Z (R_1 + h)(R_2 + h) \, d\sigma = F + 2Mh + 4\pi h^2,$$

where

$$(4.6) \qquad M = \frac{1}{2} \int_Z (R_1 + R_2) \, d\sigma = \frac{1}{2} \int_{\partial K} \left(\frac{1}{R_1} + \frac{1}{R_2} \right) df$$

is the integral of mean curvature of ∂K. If V denotes the volume of K and V_h that of K_h, from (4.5) we deduce

$$(4.7) \qquad V_h = V + \int_0^h F_h \, dh = V + Fh + Mh^2 + \tfrac{4}{3}\pi h^3,$$

which is the so-called Steiner's formula for parallel convex bodies in E_3.

For plane convex sets, the formula analogous to (4.7) is

$$(4.8) \qquad f_h = f + uh + \pi h^2,$$

where $u = $ length of ∂k. Applying (4.8) to the orthogonal projection of K on a plane perpendicular to the direction σ, we have

$$F_{\sigma, h} = F_\sigma + u_\sigma h + \pi h^2,$$

and by Cauchy's formula,

$$(4.9) \quad F_h = \frac{1}{\pi} \int_Z F_{\sigma, h} \, d\sigma = \frac{1}{\pi} \int_Z F_\sigma \, d\sigma + \frac{h}{\pi} \int_Z u_\sigma \, d\sigma + 4\pi h^2.$$

Comparing (4.9) with (4.5), we get (since both formulas hold for any h)

$$(4.10) \qquad M = \frac{1}{2\pi} \int_Z u_\sigma \, d\sigma,$$

which is a very useful expression for the integral of mean curvature of the boundary of a convex body.

On the other side, considering the volume V of K as a sum of pyramids with the common vertex O, we have

(4.11) $$V = \tfrac{1}{3} \int_{\partial K} p \, df = \tfrac{1}{3} \int_Z p R_1 R_2 \, d\sigma.$$

Applying this formula to K_h, we get

$$V_h = \frac{1}{3} \int_Z (p + h)(R_1 + h)(R_2 + h) \, d\sigma$$

$$= V + \frac{h}{3} \int_Z (R_1 R_2 + p(R_1 + R_2)) \, d\sigma$$

$$+ \frac{h^2}{3} \int_Z (p + R_1 + R_2) \, d\sigma + \frac{4}{3} \pi h^3.$$

Comparison with (4.7) yields

(4.12) $$F = \tfrac{1}{2} \int_Z p(R_1 + R_2) \, d\sigma, \qquad M = \int_Z p \, d\sigma.$$

The last formula allows definition of M for any convex body without the conditions of regularity necessary to define the principal radii of curvature of ∂K. A practical way to compute M for convex surfaces ∂K not sufficiently smooth is to compute the integral of mean curvature M_h of the parallel surface ∂K_h (which is smooth) and then to pass to the limit for $h \to 0$. This method yields the following results easily. (1) For a convex polyhedron the edges of which have lengths a_i and the corresponding dihedral angles of which are α_i, we have

$$M = \tfrac{1}{2} \sum (\pi - \alpha_i) a_i.$$

(2) For a right cylinder of height h and radius r,

$$M = \pi h + \pi^2 r.$$

(3) For a plane convex domain, considered as a flattened convex body of E_3, we have

$$M = \frac{\pi}{2} u,$$

where u is the length of the boundary of the domain.

Notice that, according to (3.22), the second formula (4.12) gives the measure of the set of planes E which cut K—that is, we have the formula

(4.13) $$\int_{E \cap K \neq 0} dE = M.$$

On the other side, applying (3.28) to convex surfaces ($n = 2$), we get

(4.14) $$\int_{G \cap K \neq 0} dG = \frac{\pi}{2} F.$$

We may therefore state:

The volume V of a convex body K is the measure of the points contained in it; the area F of ∂K is (up to the constant factor $\pi/2$) the measure of the lines which intersect K; the integral of mean curvature M is the measure of the planes which intersect K.

These integral geometric interpretations of V, F, and M have been generalized to convex bodies of the n dimensional euclidean space [(60) and Hadwiger (23) and (25)].

5. THE KINEMATIC FUNDAMENTAL FORMULA IN E_3

5.1. *The Euler characteristic of a domain.* Let Σ be a closed surface in E_3 which is of class C^2 and bounds a domain D of volume V. If df is the area element of Σ and $d\sigma$ the area element of the corresponding spherical image, we know the formulas

(5.1) $$\frac{d\sigma}{df} = \frac{1}{R_1 R_2}, \qquad I(\Sigma) = \int_\Sigma \frac{1}{R_1 R_2} df = 4\pi\chi,$$

where R_1, R_2 are the principal radii of curvature, $I(\Sigma)$ denotes the area of the spherical image of Σ, and $\chi = \chi(D)$ is the Euler characteristic of D. Because Σ is closed, its spherical image covers the unit sphere an integer number of times, and therefore $\chi = I(\Sigma)/4\pi$ is an integer. For example, for domains topologically equivalent to the solid sphere, $\chi = 1$, and for domains which are topologically equivalent to a torus, $\chi = 0$ [see, for example, Struik (69, p. 159)].

If Σ is not of class C^2 but consists of a finite number of faces (= pieces of class C^2) which intersect along edges (= closed

curves of class C^2), the Euler characteristic is obtained adding
to the area of the spherical image of the faces (5.1) the area of the
spherical image corresponding to the edges, which we shall now
compute. Let Γ be an edge of Σ and let T, N, B, denote its unit
vectors tangent, principal normal, and binormal; let s be the arc
length of Γ. If e_3, e_3' are the outward normal unit vectors to the
faces of Σ at the points of Γ and we call θ_1, θ_1' the angles which
they form with $-N$, the spherical image corresponding to Γ is
the portion of unit sphere defined by the equation,

$$Y(s, \theta) = -\cos \theta\, N + \sin \theta\, B \quad (\theta_1 \leqq \theta \leqq \theta_1', 0 \leqq s \leqq L),$$

where L is the length of Γ.

Using Frenet's formulas, we have $Y_\theta^2 = 1$, $Y_s Y_\theta = -\tau$, $Y_s^2 = \varkappa^2 \cos^2 \theta + \tau^2$, $(Y_\theta^2 Y_s^2 - (Y_s Y_\theta)^2)^{1/2} = \varkappa \cos \theta$, where $\varkappa$ and τ are
the curvature and the torsion of Γ. The area $I(\Gamma)$ of the spherical
image corresponding to Γ will be

$$(5.2) \qquad I(\Gamma) = \int_\Gamma \varkappa \cos \theta\, d\theta\, ds = \int_\Gamma (\sin \theta_1' - \sin \theta_1)\, \varkappa ds.$$

Under the assumption that Σ has no vertices (= points in
which more than two different faces intersect), the Euler characteristic of Σ is given by the second formula (5.1); we take into
account that at the left side, the integral analogous to (5.2) for
all the edges of Σ should be added.

5.2. *The kinematic formula.* Let D_0, D_1 be two domains of E_3
bounded respectively by the surfaces Σ_0, Σ_1, which we assume to
be of class C^2. Let V_i, χ_i be the volume and the Euler characteristic of D_i and let F_i, M_i be the area and the integral of mean
curvature of Σ_i $(i = 0, 1)$. Suppose D_0 is fixed and D_1 is moving,
and let dK be the kinematic density for D_1. If $\Phi(D_0 \cap D_1)$ denotes
a function of the intersection $D_0 \cap D_1$, one of the main purposes
of the integral geometry is the evaluation of integrals of the type

$$(5.3) \qquad J = \int \Phi(D_0 \cap D_1)\, dK$$

over all positions of D_1. For example, if $\Phi = V_{01} =$ volume of
$D_0 \cap D_1$, we can easily prove that $\int V_{01}\, dK = 8\pi^2 V_0 V_1$, and if

$\Phi = F_{01}$ is the area of the boundary of $D_0 \cap D_1$, the formula $\int F_{01}\, dk = 8\pi^2(V_0F_1 + V_1F_0)$ holds (50). The most important case corresponds to $\Phi = \chi(D_0 \cap D_1)$ is the Euler characteristic of $D_0 \cap D_1$. Surprisingly enough, the integral $\int \chi(D_0 \cap D_1)\, dK$ over all positions of D_1 can be expressed by only V_i, χ_i, F_i, M_i $(i = 0, 1)$. The result is the following:

$$(5.4) \quad \int \chi(D_0 \cap D_1)\, dK = 8\pi^2(V_0\, \chi_1 + V_1\chi_0) + 2\pi(F_0M_1 + F_1M_0).$$

This result is the so-called kinematic fundamental formula, which we shall now prove.

We need to compute $\chi(D_0 \cap D_1)$. The boundary of $D_0 \cap D_1$ consists in a part Σ_{01} of Σ_0 which is interior to D_1 and a part Σ_{10} of Σ_1 which is interior to D_0. Both Σ_{01} and Σ_{10} are of class C^2 and are joined by an edge $\Gamma = \Sigma_0 \cap \Sigma_1$, composed of one or more closed curves, of the boundary of $D_0 \cap D_1$. According to (5.1) we will have

$$(5.5) \qquad 4\pi\chi(D_0 \cap D_1) = I(\Sigma_{01}) + I(\Sigma_{10}) + I(\Gamma),$$

and we can write

$$(5.6) \quad 4\pi \int \chi(D_0 \cap D_1)\, dK = \int I(\Sigma_{01})\, dK \\ + \int I(\Sigma_{10})\, dK + \int I(\Gamma)\, dK,$$

where the integrals are extended over all positions of D_1.

The first two integrals on the right-hand side of (5.6) are easily evaluated. Taking the first integral, let P be a point of $\Sigma_0 \cap D_1$ and let $d\sigma_P$ denote the area element of the unit sphere at the spherical image of P. By first fixing D_1 and then letting P vary over $\Sigma_0 \cap D_1$, we get

$$\int_{P \in \Sigma_0 \cap D_1} d\sigma_P\, dK = \int I(\Sigma_{01})\, dK,$$

and by first fixing P and then rotating D_1 about this point and letting it vary over D_1 and Σ_0,

$$\int\limits_{P \in \Sigma_0 \cap D_1} d\sigma_P \, dK = \int\limits_{P \in \Sigma_0} d\sigma_P \int\limits_{P \in D_1} dK = 8\pi^2 V_1 \int\limits_{P \in \Sigma_0} d\sigma_P$$

$$= 8\pi^2 V_1 I(\Sigma_0) = 32\pi^3 V_1 \chi_0.$$

Thus, we have

$$(5.7) \qquad \int I(\Sigma_{01}) \, dK = 32\pi^3 V_1 \chi_0.$$

Similarly, by the evident invariance of the kinematic density under the inversion of the motion, we have

$$(5.8) \qquad \int I(\Sigma_{10}) \, dK = 32\pi^3 V_0 \chi_1.$$

It remains to evaluate the third integral in (5.6). Let Q be a point of Γ. By Meusnier's theorem, if ρ is the radius of curvature of Γ and R, r are the radii of normal curvature of Σ_0 and Σ_1 in the direction of the tangent to Γ at Q, we have

$$(5.9) \qquad \rho = R \cos \theta_1 = r \cos \theta_1'$$

where θ_1, θ_1' are the angles between the outward normals e_3, e_3' to Σ_0, Σ_1 at Q and the vector $-N$ opposite to the principal normal N of Γ at Q. Taking into account the identity

$$(5.10) \qquad \frac{\sin \theta_1' - \sin \theta_1}{\cos \theta_1' + \cos \theta_1} = \tan \frac{1}{2} (\theta_1' - \theta_1),$$

and putting $\theta_1' - \theta_1 = \theta$, we deduce from (5.9) and (5.10)

$$(5.11) \qquad \sin \theta_1' - \sin \theta_1 = \rho \left(\frac{1}{R} + \frac{1}{r} \right) \tan \frac{1}{2} \theta.$$

If τ_0, τ_1 denote the angles between the tangent to Γ at Q and the first principal direction of Σ_0, Σ_1 at Q, by Euler's theorem we have

$$(5.12) \qquad \frac{1}{R} = \frac{\cos^2 \tau_0}{R_1} + \frac{\sin^2 \tau_0}{R_2}, \qquad \frac{1}{r} = \frac{\cos^2 \tau_1}{r_1} + \frac{\sin^2 \tau_1}{r_2},$$

where R_1, R_2 are the principal radii of curvature of Σ_0, and r_1, r_2 are those of Σ_1 at Q. By (5.2), (5.11), and (3.33) we have

$$(5.13) \quad \int I(\Gamma) \, dK = \int \left(\frac{\cos^2 \tau_0}{R_1} + \frac{\sin^2 \tau_0}{R_2} + \frac{\cos^2 \tau_1}{r_1} + \frac{\sin^2 \tau_1}{r_2} \right)$$

$$\tan \tfrac{1}{2}\theta \, \sin^2 \theta \, df_0 \, d\tau_0 \, df_1 \, d\tau_1 \, d\theta,$$

where the limits of integration for the angles are

$$0 \leqq \tau_0 \leqq 2\pi, \qquad 0 \leqq \tau_1 \leqq 2\pi, \qquad 0 \leqq \theta \leqq \pi.$$

Computing the integral in (5.13), we get

(5.14) $$\int I(\Gamma) \, dK = 8\pi^2(F_0M_1 + F_1M_0).$$

Adding (5.7), (5.8), and (5.14), and considering (5.6), we get the desired result (5.4).

The formula (5.4) is the work of Blaschke (3). It has been generalized to E_n by Chern (8). For the generalization to spaces of constant curvature (noneuclidean geometry) see Wu (76) and (54), (57), and (58). For another kind of proof valid for more general domains than those considered here, see Hadwiger (23).

Notice that if D_0, D_1 are convex bodies, we have $\chi(D_0) = \chi(D_1) = \chi(D_0 \cap D_1) - 1$ if $D_0 \cap D_1 \not= 0$, and $\chi(D_0 \cap D_1) = 0$, if $D_0 \cap D_1 = 0$. The formula (5.4) yields

(5.15) $$\int_{D_0 \cap D_1 \neq 0} dK = 8\pi^2(V_0 + V_1) + 2\pi(F_0M_1 + F_1M_0),$$

which gives the measure of the set of congruent convex bodies D_1 having a common point with a fixed convex body D_0.

If D_1 is a sphere of radius r, we can take the origin of the moving frame at the center of D_1; then we have $\int dK = 8\pi^2 \int dP$, and (5.15) gives

$$\int dP = V_0 + F_0r + M_0r^2 + \tfrac{4}{3}\pi r^3,$$

the Steiner's formula (4.7).

6. INTEGRAL GEOMETRY IN COMPLEX SPACES

6.1. *The unitary group.* The integral geometry of complex spaces has not been developed very much, and it deserves further study. We shall give a simple typical example.

Let P_n be the n-dimensional complex projective space with the homogeneous coordinates $z_i(i = 0, 1, \cdots, n)$, so that $z = (z_0, z_1, z_2, \cdots, z_n)$ and $\lambda z = (\lambda z_0, \lambda z_1, \cdots, \lambda z_n)$, where λ is a nonzero complex number; define the same point. Let $\bar{z}_i$ denote the complex conjugate of z_i. We assume the homogeneous coordinates z_i are normalized so that

$$(6.1) \qquad (z\bar{z}) = \sum_0^n z_i \bar{z}_i = 1,$$

which determine z_i up to a factor of the form $\exp{(i\alpha)}$.

We consider the group $\mathfrak{U}$ (unitary group) of linear transformations

$$(6.2) \qquad z' = Az$$

which leaves the form (6.1) invariant. The matrices A satisfy

$$(6.3) \qquad A\overline{A}^t = E, \qquad A^{-1} = \overline{A}^t, \qquad \overline{A}^t A = E,$$

where E is the unit matrix. These relations show that $\mathfrak{U}$ depends upon $(n+1)^2$ real parameters. If we interpret the elements a_{hk} ($h = 0, 1, \cdots, n$) of the matrix A as the homogeneous coordinates of a point $a_k \in P_n$, the conditions (6.3) give

$$(6.4) \qquad (a_j \bar{a}_k) = \delta_{jk},$$

which show that the points a_k are normalized; they form the vertices of an autoconjugate n-simplex with respect to the quadric $(z\bar{z}) = 0$. Because a_k and $a_k \exp{(i\alpha_k)}$ are the same geometric point, to determine an element $u \in \mathfrak{U}$ we must give the $n + 1$ geometric points a_k [with the conditions of (6.4)], as well as the $n + 1$ real parameters α_k.

The invariant matrix of Maurer-Cartan is

$$(6.5) \qquad \omega = A^{-1} dA = \overline{A}^t dA,$$

which satisfies, in consequence of (6.3),

$$(6.6) \qquad \omega + \bar{\omega}^t = 0.$$

The invariant pfaffian forms are

$$(6.7) \qquad \omega_{jk} = \sum_{h=0}^n \bar{a}_{hj} da_{hk} = (\bar{a}_j da_k),$$

and (6.6) gives

$$(6.8) \qquad \omega_{jk} + \bar{\omega}_{kj} = 0.$$

The kinematic density of $\mathfrak{U}$, up to a constant factor, is

(6.9) $du = [\Pi\, \omega_{jk}\bar{\omega}_{jk}\, \Pi\, \omega_{hh}], \quad j < k, \quad 0 \le j, k, h \le n,$

where the product is exterior.

We have all necessary elements for the study of the integral geometry of the unitary group. We shall restrict ourselves to the case $n = 2$ (complex projective plane).

6.2. *Meromorphic curves.* A complex analytic mapping $E_1 \to P_2$ of the complex euclidean line E_1 into the complex projective plane P_2 defines a meromorphic curve in the sense of J. Weyl, H. Weyl (75), and L. Ahlfors (1); it is defined by three analytic functions $z_i = z_i(t)$, $(i = 0, 1, 2)$. Every such curve Γ has an invariant integral with respect to $\mathfrak{U}$, which we shall call the order of Γ. When the homogeneous coordinates z_i are normalized such that the condition (6.1) is satisfied, the order of Γ is defined by the following integral (up to the sign which depends upon the orientation assumed for Γ),

(6.10) $$J = \frac{1}{2\pi i}\int_\Gamma \Omega$$

where $i = \sqrt{-1}$ and

(6.11) $\Omega = [dz\, d\bar{z}] = dz_0 \wedge d\bar{z}_0 + dz_1 \wedge d\bar{z}_1 + dz_2 \wedge d\bar{z}_2.$

If Γ is an algebraic curve, we shall see that J coincides with its ordinary order or grad.

If the coordinates z_i are not normalized, we set $Z_i = z_i/(z\bar{z})^{1/2}$, and an easy calculation gives

(6.12) $$\Omega = [dZ\, d\bar{Z}] = \frac{|z \wedge z'|^2}{z^4}\ dt \wedge d\bar{t},$$

where $z \wedge z'$ denotes the vector with the components $z_1 z_2' - z_2 z_1'$, $z_2 z_0' - z_0 z_2'$, and $z_0 z_1' - z_1 z_0'$.

For some purposes, it is convenient to write Ω in another form. Let c be a point on the tangent to Γ at the point z such that

(6.13) $(c\bar{c}) = 1, \qquad (\bar{c}z) = 0.$

We will have (since c is on the tangent to Γ at z),

(6.14) $dz = \alpha z + \beta c, \qquad d\bar{z} = \bar{\alpha}\bar{z} + \bar{\beta}\bar{c}$

where α, β are the pfaffian forms

(6.15) $\alpha = (\bar{z}\, dz), \qquad \beta = (\bar{c}\, dz).$

From (6.13) and (6.14), we deduce

(6.16) $\Omega = [dz\, d\bar{z}] = \alpha \wedge \bar{\alpha} + \beta \wedge \bar{\beta},$

and because $\alpha = -\bar{\alpha}$, we have $\alpha \wedge \bar{\alpha} = 0$. Therefore,

(6.17) $\Omega = \beta \wedge \bar{\beta} = (\bar{c}\, dz) \wedge (c\, d\bar{z}),$

a formula which will be useful in the following discussion.

As an application, we shall use (6.17) to obtain the order of a complex straight line. Since J is invariant by unitary transformations and any line can be transformed into the axis $z_1 = 0$, it suffices to compute the order in this case. We take, in order to satisfy (6.1) and (6.13),

$$z = (\rho e^{i\varphi}(1 + \rho^2)^{-1/2}, 0, (1 + \rho^2)^{-1/2}),$$
$$c = (-e^{i\varphi}(1 + \rho^2)^{-1/2}, 0, \rho(1 + \rho^2)^{-1/2}),$$

and we get

$$(\bar{c}\, dz) = -\frac{d\rho + i\rho\, d\varphi}{1 + \rho^2}, \qquad (c\, d\bar{z}) = -\frac{d\rho - i\rho\, d\varphi}{1 + \rho^2}$$

and

(6.18) $\Omega = (\bar{c}\, dz) \wedge (c\, d\bar{z}) = \frac{2i\rho}{(1 + \rho^2)^2}\, d\varphi \wedge d\rho.$

The order of the segment $a \leqq \rho \leqq b$, $0 \leqq \varphi \leqq 2\pi$ will be

$$J = \frac{1}{2\pi i} \int_a^b \int_0^{2\pi} \frac{2i\rho}{(1 + \rho^2)^2}\, d\varphi\, d\rho = \frac{b^2 - a^2}{(1 + a^2)(1 + b^2)}.$$

For $a = 0$, $b = \infty$, we obtain $J = 1$, which is the order of a line.

6.3. *A generalization of the theorem of Bezout.* Let Γ_1, Γ_2 be two meromorphic curves of P_2 of orders J_1, J_2, respectively. Let $u\Gamma_2$ be the transform of Γ_2 by $u \in \mathfrak{U}$. In the theory of meromorphic curves it is important to determine the difference between the product $J_1 J_2$ and the number $N(\Gamma_1 \cap u\Gamma_2)$ of points of intersection of Γ_1 and $u\Gamma_2$, each counted with its proper multiplicity [Ahlfors (1), Chern (9) and (10), and H. Weyl (75)].

Our goal is more simple. We wish to obtain the mean value of $N(\Gamma_1 \cap u\Gamma_2)$ for all $u \in \mathfrak{U}$. First, we will compute the integral

$$(6.19) \qquad I = \int_{\mathfrak{U}} N(\Gamma_1 \cap u\Gamma_2) \, du$$

where the element of volume du is given by (6.9). In our case, $n = 2$, making use of (6.8), and considering only the absolute value, we have

$$(6.20) \quad du = (\bar{a}_0 \, da_1) \wedge (\bar{a}_1 \, da_0) \wedge (\bar{a}_0 \, da_2) \wedge (\bar{a}_2 \, da_0) \wedge (\bar{a}_1 \, da_2)$$
$$\wedge (\bar{a}_2 \, da_1) \wedge (\bar{a}_0 \, da_0) \wedge (\bar{a}_1 \, da_1) \wedge (\bar{a}_2 \, da_2).$$

Inasmuch as we are only interested in the transformations u such that $\Gamma_1 \cap u\Gamma_2 \neq 0$, we may choose the points a_0, a_1, and a_2, which determine u, so that: $a_0 = $ point of $\Gamma_1 \cap u\Gamma_2$; $a_1 = $ point on the tangent to $u\Gamma_2$ at a_0; a_2 is then determined by the relations (6.4), which we now write

$$(6.21) \quad (a_0\bar{a}_0) = (a_1\bar{a}_1) = (a_2\bar{a}_2) = 1, \; (a_0\bar{a}_1) = (a_0\bar{a}_2) = (a_1\bar{a}_2) = 0.$$

Let s be the point in which the line determined by a_1, a_2 intersects the tangent to Γ_1 at a_0. We shall have

$$(6.22) \qquad (s\bar{s}) = 1, \qquad (s\bar{a}_0) = 0, \qquad (\bar{s}a_0) = 0.$$

According to (6.17), the differential form which gives the order of $u\Gamma_2$ is

$$(6.23) \qquad \Omega_2 = (\bar{a}_1 \, da_0) \wedge (a_1 \, d\bar{a}_0) = (\bar{a}_0 \, da_1) \wedge (\bar{a}_1 \, da_0).$$

Since we always take a_0 on Γ_1, we have $da_0 = \alpha a_0 + \beta s$, where $\alpha = (\bar{a}_0 \, da_0)$, $\beta = (\bar{s} \, da_0)$. Consequently, we have

$$(\bar{a}_2 \, da_0) = \beta(\bar{a}_2 s), \qquad (a_2 \, d\bar{a}_0) = \bar{\beta}(a_2\bar{s}),$$

and, by exterior multiplication,

$$(6.24) \quad (\bar{a}_2 \, da_0) \wedge (a_2 \, d\bar{a}_0) = (\bar{a}_0 \, da_2) \wedge (\bar{a}_2 \, da_0)$$
$$= (\beta \wedge \bar{\beta})(\bar{a}_2 s)(a_2\bar{s}) = (\bar{a}_2 s)(a_2\bar{s})\Omega_1,$$

where Ω_1 is the differential form which gives the order of Γ_1.

From (6.20), (6.23), and (6.24), we have

$$(6.25) \quad du = \Omega_2 \wedge \Omega_1(\bar{a}_2 s)(a_2\bar{s}) \wedge (\bar{a}_1 \, da_2) \wedge (\bar{a}_2 \, da_1) \wedge (\bar{a}_0 \, da_0)$$
$$\wedge (\bar{a}_1 \, da_1) \wedge (\bar{a}_2 \, da_2).$$

We first keep fixed the geometric points a_0, a_1, and a_2. With the normalization (6.21), their homogeneous coordinates a_{hj} ($h = 0, 1, 2$) are determined up to an exponential factor $\exp(i\alpha_j)$; the parameters $\alpha_j(j = 0, 1, 2)$ are variables in (6.25). Putting $a_j = a_j^*$ $\exp(i\alpha_j)$, we have $da_j = a_j i\, d\alpha_j$, $(\bar{a}_j\, da_j) = i\, d\alpha_j$, and, consequently, $\int (\bar{a}_j\, da_j) = 2\pi i$ ($j = 0, 1, 2$).

From the right side of (6.25) it remains to evaluate (a_0 being fixed) $\int (\bar{a}_2 s)(a_2 \bar{s})(\bar{a}_1\, da_2) \wedge (\bar{a}_2\, da_1)$, where a_1, a_2 describe the line $(\bar{a}_0 z) = 0$ which contains the point s. We can assume, because of the invariance of the integrand by unitary transformations, that this line is the axis $z_1 = 0$. According to (6.18), we then have

$$(6.26) \qquad \int (\bar{a}_1\, da_2) \wedge (\bar{a}_2\, da_1) = \int \frac{2i\rho}{(1 + \rho^2)^2}\, d\rho\, d\varphi,$$

where we have put $a_2 = (\rho e^{i\varphi}(1 + \rho^2)^{-1/2}, 0, (1 + \rho^2)^{-1/2})$, $a_1 = (-e^{i\varphi}(1 + \rho^2)^{-1/2}, 0, \rho(1 + \rho^2)^{-1/2})$. Taking $s = (0, 0, 1)$, we obtain

$$(6.27) \qquad\qquad\qquad (\bar{a}_2 s)(a_2 \bar{s}) = \frac{1}{1 + \rho^2},$$

and, therefore,

$$(6.28) \quad \int (\bar{a}_2 s)(a_2 \bar{s})(\bar{a}_1\, da_2) \wedge (\bar{a}_2\, da_1)$$

$$= \int_0^\infty \int_0^{2\pi} \frac{2i\rho}{(1 + \rho^2)^3}\, d\rho\, d\varphi = \pi i.$$

From (6.25) and (6.28), we obtain the integral of du extended over all u such that $\Gamma_1 \cap u\Gamma_2 \neq 0$, each u counted $N(\Gamma_1 \cap u\Gamma_2)$ times. We get (up to the sign which is unessential),

$$(6.29) \qquad\qquad \int_{\mathfrak{u}} N(\Gamma_1 \cap u\Gamma_2)\, du = 32\pi^6 J_1 J_2,$$

where J_1 and J_2 are the orders of Γ_1 and Γ_2, respectively.

To obtain the mean value of $N(\Gamma_1 \cap u\Gamma_2)$, we need the total measure of $\mathfrak{U}$. Taking for Γ_1 and Γ_2 two straight lines, we know that $J_1 = J_2 = 1$ and $N = 1$; therefore (6.29) gives $\int_{\mathfrak{u}} du = 32\pi^6$. Consequently, the mean value of N is

(6.30) $$\overline{N} = J_1 J_2.$$

For algebraic curves, N is constant and (6.30) gives the classical theorem of Bezout; therefore our result may be considered a generalization of this theorem to meromorphic curves. For the extension to analytic manifolds of P_n see (56).

7. INTEGRAL GEOMETRY IN RIEMANNIAN SPACES

7.1. *Geodesics which intersect a fixed surface.* The methods of the integral geometry can be also applied to Riemannian spaces, mainly to spaces of constant curvature or other spaces which admit a group of transformations into themselves. The case of surfaces is simple and well known (55). Here, we want to consider the case of 3-dimensional spaces.

Let R_3 be a 3-dimensional Riemannian space defined by $ds^2 = g_{ij}\, dx_i\, dx_j$, where the summation convention is adopted; i, j are summed from 1 to 3. Let us introduce the notations,

$$(7.1) \qquad F = (g_{ij}x_i'x_j')^{1/2}, \qquad p_i = \frac{\partial F}{\partial x_i'},$$

where $x_i' = dx_i/dt$. As we know, a geodesic of R_3 is determined by a point x_i and a direction x_i', which is equivalent to give x_i, p_i $(i = 1, 2, 3)$. The density for sets of geodesics is defined by the following exterior differential form, taken always in absolute value:

$$(7.2) \quad dG = dp_2 \wedge dx_2 \wedge dp_3 \wedge dx_3 + dp_3 \wedge dx_3 \wedge dp_1 \wedge dx_1$$
$$+ dp_1 \wedge dx_1 \wedge dp_2 \wedge dx_2.$$

The measure of a set of geodesics is the integral of dG extended over the set. The density (7.2) is the second power of the differential invariant $\sum_1^3 dp_i \wedge dx_i$, which constitutes the invariant integral of Poincaré of the dynamics (6, pp. 19 and 78), and it therefore possesses the following two properties of invariance: (1) it is invariant with respect to a change of coordinates in the space; (2) it is invariant under displacements of the elements (x_i, p_i) on the respective geodesic.

To give a geometrical interpretation of dG, let us consider a fixed surface Σ and a set of geodesics which intersect Σ. Let G be such a geodesic and P its intersection point with Σ. In a neighborhood of P we may assume that the equation of Σ is $x_3 = 0$ and that the coordinate system is orthogonal, that is, $ds^2 = g_{11}\, dx_1^2 + g_{22}\, dx_2^2 + g_{33}\, dx_3^2$, and thus $p_i = g_{ii}(dx_i/ds)$. If ν_i represents the cosine of the angle between G and the x_i-coordinate curve at P, we have

$$(7.3) \quad \nu_i = \sqrt{g_{ii}}\,\frac{dx_i}{ds}, \quad p_i = \sqrt{g_{ii}}\,\nu_i, \quad dp_i = \sqrt{g_{ii}}\,d\nu_i + \frac{\partial \sqrt{g_{ii}}}{\partial x_h}\,\nu_i\, dx_h.$$

To determine G according to the second property of invariance of dG, we may choose its intersection point P with Σ. At this point we have $x_3 = 0$, $dx_3 = 0$, and, consequently, (7.2) takes the form

$$(7.4) \qquad\qquad dG = dp_1 \wedge dx_1 \wedge dp_2 \wedge dx_2,$$

or, according to (7.3),

$$(7.5) \qquad\qquad dG = \sqrt{g_{11}g_{22}}\,d\nu_1 \wedge dx_1 \wedge d\nu_2 \wedge dx_2.$$

On the other hand, to each set of direction cosines ν_1, ν_2, and ν_3 corresponds a point of the unit euclidean sphere and the area element in it has the value (3.15)

$$(7.6) \qquad\qquad d\sigma = \frac{d\nu_1 \wedge d\nu_2}{\nu_3}.$$

Hence, we have, in absolute value,

$$(7.7) \qquad\qquad dG = |\cos \varphi|\, d\sigma \wedge df,$$

where φ is the angle between the tangent to G and the normal to Σ at P, and $df = \sqrt{g_{11}g_{22}}\,dx_1 \wedge dx_2$ is the element of area Σ at P.

Integrating over all geodesics which intersect Σ, on the left side each geodesic is counted a number of times equal to the number n of intersection points of G and Σ; on the right, the integral of $|\cos \varphi|\, d\sigma$ gives one-half the projection of the unit sphere upon a diametral plane $(= \pi)$. Consequently, we get the integral formula

$$(7.8) \qquad\qquad \int n\, dG = \pi F,$$

where F is the area of Σ. This formula generalizes (3.28) to Riemannian spaces.

7.2. *Sets of geodesic segments.* Let t be the arc length on the geodesic G. From (7.7) we deduce

$$(7.9) \qquad dG \wedge dt = |\cos \varphi| \, d\sigma \wedge df \wedge dt.$$

The product $|\cos \varphi| \, dt$ equals the projection of the arc element dt upon the normal to Σ at P; consequently, $|\cos \varphi| \, df \wedge dt$ equals the element of volume dP of the space at P, and (7.9) can be written in the form,

$$(7.10) \qquad dG \wedge dt = dP \wedge d\sigma.$$

An oriented segment S of geodesic is determined either by G, t ($G =$ geodesic which contains S; $t =$ abscissa on G of the origin of S) or by P ($=$ origin of S) and the point of the unit euclidean sphere which gives the direction of S. The two equivalent forms (7.10) may therefore be taken as density for sets of segments of geodesic lines.

For example, let us consider the set of oriented segments S with the origin inside a fixed domain D. The integral of the left of (7.10) gives $2 \int \lambda \, dG$, where λ denotes the length of the arc of G which lies inside D (the factor 2 appears as a consequence that dG means the density for non-oriented geodesic lines). The integral of the right is equal to $4\pi V$, where V is the volume of D. Consequently, we have the following integral formula

$$(7.11) \qquad \int \lambda \, dG = 2\pi V,$$

where the integral is extended over all geodesics which intersect D.

7.3. *Some integral formulas for convex bodies in spaces of constant curvature.* Let R_3 now be a 3-dimensional space of constant curvature k. With respect to a system of geodesic polar coordinates, it is known that the element of length can be written in the form

$$(7.12) \qquad ds^2 = d\rho^2 + \frac{\sin^2 \sqrt{k}\rho}{k} \, d\tau,$$

where ρ denotes the geodesic distance from a fixed point (origin of coordinates) and $d\tau$ represents the length element of the 2-dimensional unit euclidean sphere. The volume element has the form

$$(7.13) \qquad dP = \frac{\sin^2 \sqrt{k}\rho}{k} \, d\rho \wedge d\sigma,$$

where $d\sigma$ denotes the element of area on the unit sphere.

Let P_1, P_2 be two points in R_3 such that there is only one geodesic G which unites them. Let ρ_1, ρ_2 be the abscissas on G of P_1 and P_2. With respect to a system of geodesic polar coordinates with the origin at P_1, the element of volume dP_2 has the form

$$(7.14) \qquad dP_2 = \frac{\sin^2 \sqrt{k} \, |\rho_2 - \rho_1|}{k} \, d\rho_2 \wedge d\sigma.$$

By exterior multiplication by dP_1, we have, in consequence of (7.10),

$$(7.15) \qquad dP_1 \wedge dP_2 = \frac{\sin^2 \sqrt{k} \, |\rho_2 - \rho_1|}{k} \, d\rho_1 \wedge d\rho_2 \wedge dG.$$

This formula is the work of Haimovici (27).

Let D be a convex domain of volume V (that is, it contains, with each pair of its points, the arc of geodesic, assumed unique, determined by them) and consider all the pairs P_1, P_2 inside D. The integral of the left side of (7.15) is equal to V^2. If λ denotes the length of the arc of G which lies inside D, then by calculating the integral of the right side we have

$$\int_0^\lambda \int_0^\lambda \sin^2 \sqrt{k} \, |\rho_2 - \rho_1| \, d\rho_1 \, d\rho_2 = \frac{1}{2}\left(\lambda^2 - \frac{1}{k}\sin^2 \sqrt{k}\lambda\right).$$

Hence, we have the integral formula

$$(7.16) \qquad \frac{1}{k} \int \left(\lambda^2 - \frac{1}{k}\sin^2 \sqrt{k}\lambda\right) dG = 2V^2,$$

where the integral is extended over all geodesics which intersect D.

For the elliptic space ($k = 1$), this formula reduces to

$$(7.17) \qquad \int (\lambda^2 - \sin^2 \lambda) \, dG = 2V^2,$$

and for the hyperbolic space ($k = -1$),

(7.18) $$\int (\sinh^2\lambda - \lambda^2)\, dG = 2V^2.$$

For the euclidean space $(k = 0)$, passing to the limit for $k \to 0$ in (7.16) we get

(7.19) $$\int \lambda^4\, dG = 6V^2,$$

which is a formula of Herglotz [Blaschke (3)].

Formulas of this kind referring to convex figures in the plane or to convex bodies in the euclidean space were first obtained by Crofton (7), considered the creator of the integral geometry. A great deal of them were given successively by several authors: Lebesgue (34), Blaschke (3), Massoti Biggiogero (38–42). Paper (38) contains an extensive bibliography.

The generalization to spaces of constant curvature is less known. However for certain types of formulas, the treatment in elliptic space is more satisfactory than that in euclidean space, owing to the possibility of dualization. Let us consider the following examples.

In the elliptic 3-dimensional space, all geodesics are closed and have the finite length π. The planes have finite area 2π. Since any geodesic intersects a fixed plane in one and only one point, the formula (7.8) gives the measure of the set of all geodesics of the space:

(7.20) $$\int dG = 2\pi^2.$$

Let D be a convex body of area F and volume V and let us consider the set of geodesic segments of length π which intersect D. The integral on the left of (7.10) extended over this set making use of (7.8) for $n = 2$, has the value

(7.21) $$\int dG\, dt = \pi \int dG = \frac{\pi^2}{2} F,$$

and the integral on the right is

(7.22) $$\int dP \wedge d\sigma = 2\pi V + \int_{P \notin D} \Phi\, dP,$$

where Φ denotes the solid angle under which D is seen from P

186   *L. A. Santaló*

(P exterior to D). From (7.21) and (7.22), we deduce the integral formula

$$(7.23) \qquad \int_{P \notin D} \Phi \, dP = \tfrac{1}{2}\pi^2 F - 2\pi V.$$

Let us now see which formula corresponds to (7.11) by duality. Let M, F be the integral of mean curvature and the area of the boundary of D. For the dual convex body D^* it is known that we have

$$(7.24) \quad F^* = 4\pi - F, \qquad M^* = M, \qquad V^* = \pi^2 - M - V.$$

By duality to each straight line (geodesic) G corresponds another straight line G^* and, hence, if we use (7.24), formula (7.11) gives

$$\int_{G^* \cap D^* = 0} (\pi - \varphi^*) \, dG^* = 2\pi(\pi^2 - M^* - V^*),$$

where φ^* denotes the angle between the two supporting planes of D^* through G^* and the integral is extended over all geodesics G^* exterior to D^*. Taking into account (7.20) and (7.8), and replacing G^* by G, we get the integral formula

$$(7.25) \qquad \int_{G \cap D = 0} \varphi \, DG = 2\pi(M + V) - \tfrac{1}{2}\pi^2 F,$$

which has no analogue in the euclidean geometry.

Similarly, as dual of the formula (7.17), we have

$$(7.26) \qquad \int_{G \cap D = 0} (\varphi^2 - \sin^2 \varphi) \, dG = 2(M + V)^2 - \tfrac{1}{2}\pi^3 F,$$

where, as in (7.25), φ denotes the angle between the two supporting planes of D through G and the integral is extended over all geodesics which do not intersect D. For the integral geometry in spaces of constant curvature, see Petkantschin (48), and (53), (54), and (59).

8. SUPPLEMENTARY REMARKS AND BIBLIOGRAPHICAL NOTES

8.1. *General integral geometry.* The integral geometry has its origin in the theory of geometrical probabilities [Crofton (13),

Deltheil (14), and Herglotz (29), and it was widely developed by Blaschke and his school in a series of papers quoted in Reference (3). The inclusion of the methods and results of the integral geometry within the framework of the theory of homogeneous spaces (as we have done in Section 2) is the work of Weil (73) and (74), and Chern (7). After their work, the measure theory in groups and homogeneous spaces became of fundamental interest in integral geometry. Every new result in that direction can be applied and probably exploited with success to get integral geometric statements; at least, it is sure that the integral geometry constitutes the most abundant source of examples [Nachbin (44) and Helgason (28, Chap. X)].

The inverse problem of finding a general formulation of certain particular formulas of integral geometry (Crofton's formulas) is also an interesting one [Hermann (30) Legrady (36)]. A very simple example follows. We have seen that the kinematic density for the group of motions $\mathfrak{M}$ of the plane is $dK = dP \wedge d\alpha$ (1.11). From the point of view of the homogeneous spaces, dP is the density of the space $\mathfrak{M}/\mathfrak{M}_1$, where $\mathfrak{M}_1$ denotes the group of rotations about a fixed point and $d\alpha$ is the density of $\mathfrak{M}_1$. If we write, symbolically, $dK = d\mathfrak{M}$, $dP = d(\mathfrak{M}/\mathfrak{M}_1)$, $d\alpha = d\mathfrak{M}_1$, the formula (1.11) gives $d\mathfrak{M} = d(\mathfrak{M}/\mathfrak{M}_1) \wedge d\mathfrak{M}_1$, which induces us to ask if it will hold for a general group $\mathfrak{G}$ and its subgroup $\mathfrak{g}$. In this particular example, it is well known that the formula $d\mathfrak{G} = d(\mathfrak{G}/\mathfrak{g}) \wedge d\mathfrak{g}$, in fact, holds for any locally compact topological group $\mathfrak{G}$ and any closed subgroup $\mathfrak{g}$ of $\mathfrak{G}$ [Weil (73, pp. 42–45) and Ambrose (2)].

8.2. *Sets of manifolds.* Some problems of integral geometry may also be presented under the following form. Let V denote a differentiable manifold and F a family of submanifolds in it. First we ask for the existence of a transformation group $\mathfrak{G}$ of V onto itself which transforms the elements of F onto elements of F. Then, if such a group exists, we ask for a measure of sets of varieties of F invariant under $\mathfrak{G}$. We shall give two simple examples.

Examples

1. Let V be the euclidean plane E_2 and F the family of all circles of it. The group $\mathfrak{G}$ is known to be the group of similitudes

(8.1) $x' = \rho(x \cos \varphi - y \sin \varphi) + a,$

$$y' = \rho(x \sin \varphi + y \cos \varphi) + b,$$

which depends on the 4 parameters a, b, ρ, and φ. This group can be represented by the group of matrices,

$$u = \begin{pmatrix} \rho \cos \varphi & -\rho \sin \varphi & a \\ \rho \sin \varphi & \rho \cos \varphi & b \\ 0 & 0 & 1 \end{pmatrix},$$

and by the method of Section (2.2), we find immediately that the forms of Maurer-Cartan are

$$\omega_1 = \frac{d\rho}{\rho}, \qquad \omega_2 = d\varphi, \qquad \omega_3 = \frac{\cos \varphi}{\rho} da + \frac{\sin \varphi}{\rho} db,$$

$$\omega_4 = \frac{\sin \varphi}{\rho} da + \frac{\cos \varphi}{\rho} db.$$

The similitudes which leave invariant a given circle are characterized by $a, b, \rho =$ constants, and, consequently, the system (2.3) is $\omega_1 = 0$, $\omega_3 = 0$, $\omega_4 = 0$. The density for sets of circles (of center a, b and radius ρ) invariant under the group of similitudes results:

$$dC = \frac{da \wedge db \wedge d\rho}{\rho^3}.$$

2. Let V be the real projective plane and F the family of nondegenerate conics in it. Then the group G is the projective group and the density for conics is (61),

$$dC = \frac{da_{00} \wedge da_{01} \wedge da_{02} \wedge da_{11} \wedge da_{12}}{3\Delta^2}$$

where $\Delta = \det (a_{ij})$ and the equation of the conic is assumed to be

$$a_{00}x_0^2 + 2a_{01}xy + a_{11}y^2 + 2a_{02}x + 2a_{12}y + 1 = 0.$$

Other examples of this kind have been given by Stoka (63–68). For sets of degenerate conics, see Luccioni (37).

8.3. *Integral geometry of special groups.* The metric (euclidean and noneuclidean) integral geometry is the best known; however, other cases have also been investigated. The integral geometry of the unimodular affine group of the euclidean space onto itself leads to certain affine invariants for convex bodies (62). The integral geometry of the projective group has been considered by Varga (70) and is pursued in (55); that of the symplectic group has been studied by Legrady (35).

In the last years, Gelfand and his school have largely generalized the ideas of the integral geometry and used them in problems of group representation (21).

REFERENCES

The following list contains almost exclusively the papers mentioned in the text. References (3), (23), (38) contain a more complete bibliography.

1. Ahlfors, L. V., "The theory of meromorphic curves," *Acta Soc. Sc. Fenn.*, A-III (1941), No. 4, 1–31.

2. Ambrose, W., "Direct sum theorem for Haar measures," *Transactions of the American Mathematical Society*, 61 (1947), 122–27.

3. Blaschke, W., *Vorlesungen über Integralgeometrie*, 3rd ed. Berlin: 1955.

4. Bonnesen, T., and W. Fenchel, *Theorie der konvexen Körper*. Berlin: Ergebnisse der Math., 1934.

5. Busemann, H., *Convex surfaces*. New York: Interscience, 1958.

6. Cartan, É., *Leçons sur les invariants intégraux*. Paris: Hermann, 1922.

7. Chern, S. S., "On integral geometry in Klein spaces," *Ann. of Math.*, 43 (1942), 178–89.

8. ———, "On the kinematic formula in the euclidean space of n dimensions," *American Journal of Mathematics*, 74 (1952), 227–36.

9. ———, "Differential geometry and integral geometry," *Proceedings of the International Congress of Mathematics*, Edinburgh (1958), 441–49.

10. ———, "The integrated form of the first main theorem for complex analytic mappings in several complex variables," *Ann. Math.*, 71 (1960), 536–51.

11. Chern, S. S., and R. K. Lashof, "On the total curvature of immersed manifolds," I, *American Journal of Mathematics*, 79 (1957), 306–18; II, *Michigan Mathematical Journal*, 5 (1958), 5–12.

12. Chevalley, C., *Theory of Lie Groups*. Princeton, N.J.: Princeton University Press, 1946.

13. Crofton, M. W., "On the theory of local probability," *Phil. Trans. R. Soc. London*, 158 (1868), 181–99.

14. Deltheil, R., *Probabilités géométriques*. Paris: Albin Michel, 1926.

15. Fáry, I., "Functionals related to mixed volumes," *Illinois Journal of Mathematics*, 5 (1961), 425–30.

16. ———, "Sur la courbure totale d'une courbe gauche faisant un noeud," *Bull. Soc. Math. France*, 77 (1949), 128–38.

17. Federer, H., "Coincidence functions and their integrals," *Transactions of the American Mathematical Society*, 59 (1946), 441–66.

18. ———, "The (φ, k)-rectifiable subsets of n space," *Transactions of the American Mathematical Society*, 62 (1947), 114–92.

19. ———, "Some integral geometric theorems," *Transactions of the American Mathematical Society*, 77 (1954), 238–61.

20. Fenchel, W., "On the differential geometry of closed space curves," *Bulletin of the American Mathematical Society*, 57 (1951), 44–54.

21. Gelfand, I. M., "Integral geometry and its relations to the theory of representations," *Uspehi Mat. Nauk*, 15 (1960), 155–64.

22. Green, L. W., "Proof of Blaschke's sphere conjecture," *Bulletin of the American Mathematical Society*, 67 (1961), 156–58.

23. Hadwiger, H., *Vorlesungen über Inhalt, Oberflache und Isoperimetrie*. Berlin: Springer, 1957.

24. ———, *Altes und neues über konvexe Körper*. Basel and Stuttgart: Birkhauser Verlag, 1955.

25. ———, "Normale Körper im Euklidischen Raum und ihre topologischen und metrischen Eigenschaften," *Math. Zeits.*, 71 (1959), 124–40.

26. Haimovici, M., "Géométrie intégrale sur les surfaces courbes," *C. R. Acad. Sc. Paris*, 203 (1936), 230–32.

27. ———, "Généralisation d'une formule de Crofton dans un espace de Riemann à n dimensions," *C. R. Acad. Sc. Roumanie*, 1 (1936), 291–96.

28. Helgason, S., *Differential Geometry and Symmetric Spaces*. New York: Academic Press Inc., 1962.

29. Herglotz, G., *Lectures on Geometric Probabilities* (mimeographed notes). Göttingen: 1933.

30. Hermann, R., "Remarks on the foundation of integral geometry," *Rend. Circ. Mat. Palermo, Ser. II*, 9 (1960), 91–96.

31. Kurita, M., "On the volume in homogeneous spaces," *Nagoya Math. J.*, 15 (1959), 201–17.

32. ———, "An extension of Poincaré formula in integral geometry," *Nagoya Math. J.*, 2 (1951), 55–61.

33. ———, *Integral Geometry*. Nagoya, Japan: Nagoya University, 1956.

34. Lebesgue, H., "Exposition d'un mémoire de M. W. Crofton," *Nouvelles Ann. de Math. S. IV*, 12 (1912), 481–502.

35. Legrady, K., "Symplektische Integralgeometrie," *Annali di Mat. Serie IV*, 41 (1956), 139–59.

36. ———, "Sobre la determinación de funcionales en geometría integral," *Rev. Union Mat. Argentina*, 19 (1960), 175–78.

37. Luccioni, R. E., "Sobre la existencia de medida para hipercuadricas singulares en espacios proyectivos," *Rev. Mat. y Fis. Univ. Tucuman*, 14 (1962), 269–76.

38. Masotti Biggiogero, G., "La geometria integrale," *Rend. Sem. Mat. e Fis. Milano* 25 (1953–54), 3–70.

39. ———, "Nuove formule di geometria integrale relative agli ovaloidi," *Rend. Ist. Lombardo* A 96 (1962), 666–85.

40. ———, "Nuove formule di geometria integrale relative agli ovali," *Annali di Mat.*, 4 58 (1962), 85–108.

41. ———, "Su alcune formule di geometria integrale," *Rend. di Mat. (5)* 14 (1955), 280–88.

42. ———, "Sulla geometria integrale: nuove formule relative agli ovaloidi," Scritti Mat. in onore di Filippo Sibirani, Bologna: Cesari Zuffi (1957), 173–79.

43. Munroe, M. E., *Modern Multidimensional Calculus*. Reading, Mass.: Addison-Wesley Publishing Co., Inc., 1963.

44. Nachbin, L., *Integral de Haar, Textos de Matematica*. Recife, Brazil: 1960.

45. Nöbeling, G., "Über die Flachenmasse im Euklidischen Raum," *Math. Ann.*, 118 (1943), 687–701.

46. ———, "Über den Flacheninhalt dehnungsbeschränkten Flächen," *Math. Z.*, 48 (1943), 747–71.

47. Owens, O. G., "The integral geometric definition of arc length for two-dimensional Finsler spaces," *Transactions of the American Mathematical Society*, 73 (1952), 198–210.

48. Petkantschin, B., "Zusammenhänge zwischen den Dichten der linearen Unterräume im *n*-dimensionales Raum," *Abh. Math. Sem. Univ. Hamburg* 11 (1936), 249–310.

49. Rohde, H., "Unitäre Integralgeometrie," *Abh. Math. Sem. Univ. Hamburg*, 13 (1940), 295–318.

50. Santalo, L. A., "Über das kinematische Mass im Raum," *Actualités Sci. Ind.* No. 357. Paris: Hermann, 1936.

51. ———, "A theorem and an inequality referring to rectifiable curves," *American Journal of Mathematics*, 63 (1941), 635–44.

52. ———, "Unas formulas integrales y una definicion de area *g*-dimensional de un conjunto de puntos," *Rev. de Mat. y Fis. Univ. Tucuman*, 7 (1950), 271-82.

53. ———, "Measure of sets of geodesics in a Riemannian space and applications to integral formulas in elliptic and hyperbolic spaces," *Summa Brasil Math.*, 3 (1952), 1–11.

54. ———, "Geometría integral en espacios de curvatura constante," *Com. E. Atom. Buenos Aires*, No. 1 (1952).

55. ———, "Introduction to integral geometry," *Actualités Sci. Ind.*, No. 1198. Paris: Hermann, 1953.

56. ———, "Integral geometry in Hermitian spaces," *American Journal of Mathematics*, 74 (1952), 423–34.

57. ———, "On the kinematic formula in spaces of constant curvature," *Proceedings of the International Congress of Mathematics, Amsterdam*, 2 (1954), 251–52.

58. ———, "Sobre la formula fundamental cinematica de la geometría integral en espacios de curvatura constante," *Math. Notae*, 18 (1962), 79–94.

59. ———, "Cuestiones de geometría diferencial y integral en espacios de curvatura constante," *Rend. Sem. Mat. Torino*, 14 (1954–55), 277–95.

60. ———, "Sur la mesure des espaces linéaires que coupent un corps convexe et problèmes qui s'y rattachent," *Colloque sur les questions de réalité en géométrie*, Liège (1956), 177–90.

61. ———, "Two applications of the integral geometry in affine and projective spaces," *Publ. Math. Debrecen,* 7 (1960), 226–37.

62. ———, "On the measure of sets of parallel subspaces," *Canadian Journal of Mathematics,* 14 (1962), 313–19.

63. Stoka, M., "Masure unei multimi de varietati dintr-un spatiu R_n," *Bull. Math. Soc. Sci. Ac. R. P. R.,* 7 (1955), 903–37.

64. ———, "Asupra grupurilor G_r masurabili dintri-un spatiu R_n," *Com. Acad. R. P. R.,* 7 (1957), 581–85.

65. ———, "Geometria integrale in uno spazio euclideo E_n," *Boll. Un. Mat. Italiana,* 13 (1958), 470–85.

66. ———, "Famiglie di varieta misurabili in uno spazio E_n," *Rend. Circ. Mat. Palermo, II,* 8 (1959), 1–14.

67. ———, "Géométrie intégrale dans un espace E_n," *Acad. R. P. R. IV,* (1959), 123–56.

68. ———, "Integralgeometrie in einen Riemannschem Raum V_n," *Acad. R. P. R. V* (1960), 107–20.

69. Struik, D. J., "Lectures on classical differential geometry," Cambridge, Mass.: Addison-Wesley, 1950.

70. Varga, O., "Über Masse von Paaren linearer Mannigfaltigkeiten im projektiven Raum, P_n," *Rev. Mat. Hispano-Americana,* (1935), 241–278.

71. Vidal Abascal, E., "Sobre algunos problemas en relacion con la medida en espacios foliados," *Primer Coloquio de Geometría Diferencial, Santiago de Compostela,* (1963), 63–82.

72. Vincensini, P., "Corps convexes, séries linéaires, domaines vectoriels," *Mémorial des Sciences Mathématiques,* No. 94. Paris: 1938.

73. Weil, A., "L'intégration dans les groupes topologiques et ses applications," *Actualités Sci. Ind.* No. 869. Paris: Hermann, 1940.

74. ———, Review of the paper 7 of Chern, *Mathematical Reviews,* 3 (1942), 253.

75. Weyl, H., "Meromorphic functions and analytic curves," Princeton University Press (1943).

76. Wu, T. J., "Über elliptische Geometrie," *Math. Z.,* 43 (1938), 212–27.

INDEX

absolutely continuous, 135
admissible set, 125
area, 34, 39, 160, 166, 171
   integral, 139
atlas, 70

Bernstein's theorem, 53
Betti number, 11
Bezout's theorem, 178
Bonnet's theorem, 108
bordant, 4

chain, 13
   p-chains, 74
characteristics, 8, 10
chart, 70
Christoffel symbols, 40
circle, 76
closed, 17, 20
   forms, 75
Codazzi equations, 47
Cohn-Vossen theorem, 46
cohomology group, 76
col, 9
complex variable, 7
composition mapping, 141
conjugate, 104, 106
   point, 96, 108
connection form, 40
continuous mapping, 135
continuum, 142
convex, 17, 21
   bodies, 168, 171

covariant differential, 40
   differentiations, 105
critical point, 11, 13
Crofton's theorem, 33
curvature, 17, 108
   constant, 183
   Gaussian, 30, 80
   total, 29, 32, 35, 89
cut:
   locus, 99, 100
   point, 96, 100
cycle, 10
cyclic element, 144
cylinder, 76

deformation, 36
density, 148, 152, 154, 161, 162
   kinematic, 150, 155, 164
de Rham theorem, 76
differential:
   equations, 8
   form, 71
   formula, 165
   geometry, 6
directed lines, 69

ellipsoid, 102
equilibrium:
   point, 11, 13
   theory, 10
Euler characteristic, 43, 171
Euler-Poincaré characteristic, 89

195